With Best Wishes

Richard

MODERN ASSAYS FOR PLANT PATHOGENIC FUNGI

Identification, Detection and Quantification

MODERN ASSAYS FOR PLANT PATHOGENIC FUNGI

Identification, Detection and Quantification

Edited by

A. SCHOTS

Laboratory for Monoclonal Antibodies, Wageningen, The Netherlands

F. M. DEWEY

Department of Plant Sciences, University of Oxford, UK

AND

R. P. OLIVER

School of Biological Sciences, University of East Anglia, UK

CAB INTERNATIONAL

CAB INTERNATIONAL
Wallingford
Oxford OX10 8DE
UK

Tel: Wallingford (0491) 832111
Telex: 847964 (COMAGG G)
Telecom Gold/Dialcom: 84: CAU001
Fax: (0491) 833508

A catalogue entry for this book is available from the British Library.

ISBN 0 85198 870 9

Typeset by Create Publishing Services Ltd, Bath
Printed and bound in Great Britain at the University Press, Cambridge

Contents

Introduction   ix

Part One: Stem-based Diseases

1   Molecular Diagnosis of the Rapeseed Pathogen *Leptosphaeria maculans*
     Based on RAPD-PCR   1
     C. Schäfer and J. Wöstemeyer

2   A Monoclonal Antibody-based Immunoassay for the Detection of the
     Eyespot Pathogen of Cereals *Pseudocercosporella herpotrichoides*   9
     F. M. Dewey and R. A. Priestley

3   DNA Probe for R-Type of Eyespot Disease of Cereals *Pseudocercosporella
     herpotrichoides*   17
     P. Nicholson and H. N. Rezanoor

Part Two: Soil-borne Diseases

4   Detection of *Spongospora subterranea* by ELISA Using Monoclonal
     Antibodies   23
     J. G. Harrison, R. Lowe, A. Wallace and N. A. Williams

5   Development of Monoclonal Antibody-based Immunological Assays for
     the Detection of Live Propagules of *Rhizoctonia solani* in the Soil   29
     C. R. Thornton, F. M. Dewey and C. A. Gilligan

6   Differentiation Between *Fusarium culmorum* and *Fusarium graminearum* by
     RFLP and with Species-specific DNA Probes   37
     B. Koopmann, P. Karlovsky and G. Wolf

7   RAPDs of *Fusarium culmorum* and *F. graminearum*: Application for
     Genotyping and Species Identification   47
     A. G. SCHILLING, E. M. MÖLLER AND H. H. GEIGER

Part Three: Vascular Pathogens

8   Electrophoretic Detection Method for *Fusarium oxysporum* Species in
     Cyclamen and Carnation by Using Isoenzymes   57
     A. KERSSIES, A. EVERINK, L. HORNSTRA AND H. J. VAN TELGEN

9   Detection and Identification of *Fusarium oxysporum* f.sp. *gladioli* by RFLP
     and RAPD Analysis   63
     J. J. MES, J. VAN DOORN, E. J. A. ROEBROECK AND P. M. BOONEKAMP

10   RFLPs for Distinguishing Races of the Pea Vascular Wilt Pathogen
     Fusarium oxysporum f.sp. *pisi*   69
     A. CODDINGTON, B. G. LEWIS AND D. S. WHITEHEAD

11   The Use of Amplified Fragment Length Polymorphisms (AFLPs) of
     Genomic DNA for the Characterization of Isolates of
     Fusarium oxysporum f.sp. *ciceris*   75
     A. KELLY, A. R. ALCALA-JIMÉNEZ, B. W. BAINBRIDGE, J. B. HEALE,
     E. PEREZ-ARTES AND R. M. JIMÉNEZ-DÍAZ

12   PCR-based Assays for the Detection and Quantification of *Verticillium*
     Species in Potato   83
     J. ROBB, X. HU, H. PLATT AND R. NAZAR

13   Detection and Differentiation by PCR of Subspecific Groups Within
     Two *Verticillium* Species Causing Vascular Wilts in Herbaceous Hosts   91
     J. H. CARDER, A. MORTON, A. M. TABRETT AND D. J. BARBARA

14   A Double (Monoclonal) Antibody Sandwich ELISA for the Detection
     of *Verticillium* Species in Roses   99
     M. M. VAN DE KOPPEL AND A. SCHOTS

Part Four: Root Rot and Root-infecting Fungi

15   Detection of *Heterobasidion annosum* Using Monoclonal Antibodies   105
     D. N. GALBRAITH AND J. W. PALFREYMAN

16   Identification of *Heterobasidion annosum* (Fr.) Bref., a Root and Butt Rot
     Disease of Trees, by PCR Fingerprinting   111
     R. KARJALAINEN AND K. KAMMIOVIRTA

17   The Development of Rapid Identification of *Heterobasidion annosum*
     Intersterility Groups by Ribosomal RNA Gene Polymorphism and by
     Minisatellite Allele Polymorphism   117
     T. KASUGA AND K. R. MITCHELSON

18 Identification of Fungi in the *Gaeumannomyces–Phialophora* Complex
Associated with Take-all of Cereals and Grasses Using DNA Probes 127
E. WARD AND G. L. BATEMAN

19 Detection of *Phytophthora fragariae* var. *rubi* in Infected Raspberry
Roots by PCR 135
G. STAMMLER AND E. SEEMÜLLER

20 The Detection and Species Identification of African *Armillaria* 141
C. MOHAMMED

21 The Development of Monoclonal Antibody-based ELISA and Dipstick
Assays for the Detection and Identification of *Armillaria* Species in
Infected Wood 149
R. PRIESTLEY, C. MOHAMMED AND F. M. DEWEY

Part Five: Postharvest Diseases

22 Development of an ELISA and Dot-blot Assay to Detect *Mucor racemosus*
and Related Species of the Order Mucorales 157
G. A. DE RUITER, W. BOS, A. W. VAN BRUGGEN-VAN DER LUGT, H.
HOFSTRA AND F. M. ROMBOUTS

23 Monoclonal Antibody-based ELISA for Detection of Mycelial Antigens of
Botrytis cinerea in Fruits and Vegetables 165
R. BOSSI, L. COLE, A. D. SPIER AND F. M. DEWEY

24 Monoclonal Antibodies for Detection of Conidia of *Botrytis cinerea* on Cut
Flowers Using an Immunofluorescence Assay 173
J. SALINAS, G. SCHOBER AND A. SCHOTS

25 Strawberry Blackspot Disease (*Colletotrichum acutatum*) 179
I. BARKER, G. BREWER, R. T. A. COOK, S. CROSSLEY AND S. FREEMAN

26 Detection of the Anthracnose Pathogen *Colletotrichum* 183
P. R. MILLS, S. SREENIVASAPRASAD AND A. E. BROWN

Part Six: Seed-borne Diseases

27 Development of Species-specific Probes for Identification of *Pyrenophora
graminea* and *P. teres* by Dot-blot or RFLP 191
K. HUSTED

28 Monoclonal Antibodies for the Detection of *Pyrenophora graminea* 199
R. BURNS, M. L. VERNON AND E. L. GEORGE

29 PCR for the Detection of *Pyrenophora* Species, *Fusarium moniliforme,
Stenocarpella maydis*, and the *Phomopsis/Diaporthe* Complex 205
E. J. A. BLAKEMORE, D. S. JACCOUD FILHO AND J. C. REEVES

Part Seven: Foliar Diseases

30   PCR for Detection of the Fungi that Cause Sigatoka Leaf Spots of
Banana and Plantain   215
A. JOHANSON

31   DNA Fingerprinting of the Potato Late Blight Fungus, *Phytophthora
infestans*   223
A. DRENTH AND F. GOVERS

32   Specificity of Monoclonal Antibodies Raised to *Pythium aphanidermatum*
and *Erysiphe pisi*   231
A. J. MITCHELL, A. J. MACKIE, A. M. ROBERTS, K. A. HUTCHISON,
M. T. ESTRADA-GARCIA, J. A. CALLOW AND J. R. GREEN

Part Eight: Detection of Fungicide Resistance

33   Allele-specific Oligonucleotides and their Use in Characterization of
Resistance to Benomyl in *Venturia inaequalis*   239
H. KOENRAADT AND A. L. JONES

Part Nine: Field Detection of Airborne Fungi

34   The Development of Immunological Techniques for the Detection and
Evaluation of Fungal Disease Inoculum in Oilseed Rape Crops   247
D. SCHMECHEL, H. A. McCARTNEY AND K. HALSEY

Index   255

Introduction

Detection and identification of plant pathogenic fungi is increasing in importance as the demand for pathogen-free propagation and planting material multiplies. This is mainly the result of the need to reduce the number and amount of crop protection chemicals. Until about a decade ago, the development of detection and identification methods was limited to classical methods requiring hard-won skills and taxonomic knowledge. With the advent of monoclonal antibody technology (Kohler and Milstein, 1975) and DNA fingerprinting technology, tools have become available to develop tests based on unequivocal parameters such as species-specific molecules or DNA sequences.

In 1987, an initiative was undertaken to assemble knowledge and exchange experience within the European Community, on the development and use of detection methods based on monoclonal antibodies or DNA probes. From this initiative, known as the COST-88 programme, workshops have been organized on the detection of plant pathogenic viruses, bacteria and fungi. After the 1990 workshop on plant pathogenic fungi in Dundee, Scotland, two working groups were formed to explore detection of these pathogens using monoclonal antibodies and DNA technology. From the meetings of these two working groups it became apparent that the developments in both fields were rapid. More and more assays based on specific serological and nucleic acid entities were developed or are under development. The working group therefore decided to organize a joint conference on 'Modern detection methods for plant pathogenic fungi' to disseminate this knowledge to researchers, policy makers and representatives from inspection services and plant protection agencies. The contributions to this book were presented at this conference and comprise with one exception, protocols of serological assays based on monoclonal antibodies or assays based on DNA techniques.

The question arises: 'What are the perspectives for detection and identification of fungal pathogens?'

Considerable advances have been made in the past two years in the development of immunoassays for fungi. Undoubtedly this is due to the more widespread use of hybridoma technology which has enabled researchers to raise monoclonal antibodies that are isolate-, species- and genus-specific for particular pathogens but more importantly do not cross-react with host plant material. However, some problems still remain: raising monoclonal antibodies that are specific and suitable for detection purposes is still an inefficient process as is often the extraction of fungal antigens from infected material.

No one source or type of fungal antigen has been identified as a source of immunogens that will produce a high ratio of hybridomas secreting monoclonal antibodies that are highly specific. The yield from fusions with splenocytes from mice immunized with extracts from whole mycelial or mycelial fragments is very low. Dewey and co-workers (see Dewey, 1992) have shown that saline surface washings of solid slant cultures are a good source of antigens for immunization, because antigens recognized by monoclonal antibodies arising from such fusions are easy to extract from infected plant material. However, surface washings have not proved to be a suitable source of immunogens in all cases. Removal of the immunodominant non-specific molecules from fungal washes or mycelial extracts has been shown to be helpful (Bossi and Dewey 1992; this volume, Chapter 2).

Only a few reports are available on the use of partially purified proteins as a source of antigen for immunization (Chapter 24). Now that more of the pathogenicity process is known and more and more fungi are characterized biochemically, this route should lead to more specific antibodies directed, for instance, against pathogenicity factors which are used by the fungus for colonization of the host. Fungal enzymes that degrade the host cell wall are among the antigens of choice. It is critical when such methods are used that the final assay be considered, particularly the expression and extractability of the target molecules from the host. For example, it is fortunate that in the soil detection assay for *Rhizoctonia solani*, developed by Thornton *et al.* (Chapter 5), in which mice were immunized with mycelial extracts from the fungus grown on potato starch broth, that the sensitivity and detection of live propagules were more important criteria than the development of a rapid assay. The selected monoclonal antibody recognizes an antigen, not normally present in the soil, that is only produced when the fungus is grown on potato starch broth. By incorporating a 48 h enrichment period, in which soil samples are incubated in potato broth, the assay is sensitive and detects only live propagules. De Ruiter *et al.* (Chapter 22) using methods developed by Notermans's group (see Notermans and Heuvelmann, 1985) found that high molecular weight carbohydrate extracts from fungal mycelia are a good source of immunogens for raising genus-specific antigens and Green *et al.* (unpublished) in the development of highly specific antibodies for detection of infection structures of *Colletotrichum lindemuthianum* have shown that immunization of mice with antiserum raised to host antigens blocks the production of antibodies that cross-react with host molecules.

The ease and efficiency with which fungal antigens can be extracted from infected plant material appears to vary from one pathogen/host combination to another. For example, antigens of *Heterobasidion annosum* can be detected easily in infected wood material (see Galbraith and Palfreyman, Chapter 15) but difficulty was encountered in extracting *Armillaria* antigens from *Armillaria* growing on naturally infected wood/bark samples (Priestley *et al.*, Chapter 21) and in extraction of antigens from the stem-base region of young wheat plants infected with *Pseudocercosporella herpotrichoides* (Dewey and Priestley, Chapter 2). However, with newer techniques being constantly developed to increase sensitivity, such as the electrorotation assay (Coghlan, 1993) where one specific molecule in 10^{15} can be detected by antibodies bound to magnetic spheres exposed to a rotating magnetic field, it would seem that the efficiency and sensitivity of immunoassays is almost limitless.

The development of *in vitro* technology whereby antibody genes are expressed in bacteria, will negate the need for animal house facilities and promises to make the development of monoclonal antibody immunoassays more widely accessible. (This is the subject of a complementary COST-88 working group.) Furthermore, once antibody engineering has become routine it is conceivable that researchers will make designer antibodies that will enable greater levels of specificity to be achieved.

In the future it is probable that many fungal immunoassays will be adapted for field-use as dip-sticks or dot-blots that can be used on site by the grower or adviser, similar to those developed by the Agri-Diagnostics Company for fungal diseases of turf grass. Such systems are likely to be incorporated into effective management pathogen/pest control systems and will help in effective targeting of fungicides.

There has been very rapid development of nucleic-acid-based techniques. In particular, methods based on the polymerase chain reaction (Saiki *et al.*, 1988) have taken centre stage. While restriction fragment length polymorphism (RFLP) techniques will continue to be used widely in research settings, there was widespread agreement at the conference that they have little future in routine diagnostic applications. Contributions to the meeting and this book demonstrate that the polymerase chain reaction (PCR) can live up to its billing as very sensitive, very specific and rapid. For example, Stammler and Seemüller (Chapter 19) and Mills *et al.* (Chapter 26) show that PCR can detect specific fungi when only femtogram quantities of DNA are present. These levels of sensitivity and specificity approach the theoretical maxima of detection of one nucleus of a specific genotype against a background of contaminating DNA (whether fungal or plant) that need differ at only one base. In addition, PCR assays can accurately quantify fungal infection (Robb *et al.*, Chapter 12). These bravura demonstrations have shown unequivocally that PCR techniques possess all the theoretical attributes of ideal detection systems.

Many of the primers have been developed from the rDNA sequences. This is very convenient because the ITS regions seem species-specific and yet they can easily be generated using conserved primers to the rDNA genes (see Chapters 12 and 26). One by-product of this is that unequivocal taxonomic information that is directly comparable with other fungal species is obtained. It is to be hoped that the sequences to these regions will be freely shared.

That is not to say that there are no problems. These fall into two classes. Firstly there are technical problems with PCR amplifications. These include false positives created by contaminating target DNA, where the cure appears to be renewed laboratory hygiene, and false negatives caused primarily by inhibitors in DNA preparations of fungal or infected plant material, where the cure is partly the incorporation of suitable controls. There is still a clear need for rapid development in quick, clean and easy DNA preparation techniques, e.g. as propounded by Carder *et al.* (Chapter 13). Some of these technical problems may be easier to solve when the assay is based on the ligation chain reaction (Iovannisci and Winn-Deen, 1993).

The second problem concerns the development of usable assays from PCR amplifications. PCR generates a specific double-stranded DNA product which is traditionally detected by agarose gel electrophoresis. The information content can reside in whether a band is produced, the size of the band or the sequence of the band as revealed either by restriction endonuclease digestion or even by sequencing. For practical assays, one imagines that the production of a band will constitute the assay. In this case agarose gel electrophoresis would be too slow and cumbersome for routine, automated use. The challenge then is to detect the presence of the amplimer directly from the PCR reaction. This requires the use of a 'capture reagent' which could be either (one of) the dNTPs (nucleoside triphosphates) or the primers. These can be labelled with biotin with operator sequences or more simply by UV fixation to plastic dishes (Nagata *et al.*, 1985). The second requirement is for a 'detection reagent'. This could be digoxigenin, a fluorescent tag, or radioactivity. Again these can be attached either to the primers or to the dNTPs. After the PCR the double-stranded product is captured (perhaps by use of magnetic beads attached to avidin or by DNA binding probes), unbound nucleotides and primers are washed away and the detection reagent is revealed, either by fluorescence or by enzyme-linked anti-dioxygenin antibodies (Vliet *et al.*, 1993). The permutations are endless. These automations of PCR detections are already being applied to clinical studies and we, in the plant field, stand to benefit from the development work being carried out.

It seems to us that the purely scientific problems associated with fungal phytodiagnosis have been shown to be soluble, even if not yet solved. The requirement is now for the regulatory authorities to seize the opportunities and to stimulate the adoption of these faster, more sensitive and more specific assays. These developments could then be used to reduce use of pesticides in agriculture with consequently fewer residues in food products.

REFERENCES

Bossi, R. and Dewey, F. M. (1992) Development of a monoclonal-antibody-immunodetection assay for *Botrytis cinerea* (Pers). *Plant Pathology* 41, 472–482.

Coghlan, A. (1993) Sticky beads put bacteria into a spin. *New Scientist* 1873, 21.

Dewey, F. M. (1992) Detection of plant invading fungi. In: Duncan, J. M. and Torrance, L. (eds) *Techniques for Rapid Detection of Plant Diseases*. Blackwell Scientific, Oxford, pp. 47–63.

Iovannisci, D. M. and Winn-Deen, F. S. (1993) Ligation amplification and fluorescence detection of *Mycobacterium tuberculosis*. *Molecular and Cellular Probes* 7, 35–44.

Köhler, G. and Milstein, C. (1975) Continuous cultures of fused cells secreting antibody of predefined specificity. *Nature* 256, 495.

Nagata, Y., Yokota, H., Kosuda, O., Yokoo, K., Takemura, K. and Kikuchi, T. (1985) Quantification of picogram levels of specific DNA immobilised in microtiter wells. *FEBS Letters* 183, 379–382.

Notermans, S. and Heuvelmann, C. J. (1985) Immunological detection of moulds in food by using enzyme-linked immunosorbent assay (ELISA); preparation of antigens. *International Journal of Food Microbiology* 2, 247–258.

Saiki, R. K., Gelfrand, D. I. I., Stoffel, S., Scharf, S. J., Higuchi, R., Horn, G. T., Mullis, K. B. and Erhlich, H. A. (1988) Primer-directed enzymatic amplification of DNA with a thermostable DNA-polymerase. *Science*, 240, 239–487.

Vliet, van der G. M. F., Wit, de M. Y. L. and Klaster, P. R. (1993) A simple colorimetric assay for detection of amplified *Mycobacterium leprae* DNA. *Molecular and Cellular Probes* 7, 61–66.

1

Molecular Diagnosis of the Rapeseed Pathogen *Leptosphaeria maculans* Based on RAPD-PCR

C. SCHÄFER[1] AND J. WÖSTEMEYER[2]

[1] *Institut für Gerbiologische Forschung, Ihnestrasse 63, D-14195 Berlin 33, Germany;*
[2] *Friedrich-Schiller Universität, Lehrstuhl für Mikrobiologie und Mikrobengenetik, Neugasse 24, D-07743 Jena, Germany.*

BACKGROUND

The ascomycetous fungus *Leptosphaeria maculans* and its anamorph, *Phoma lingam*, is the causative agent of blackleg disease of many crucifers. Especially in rainy years, the disease causes considerable losses in oilseed rape (*Brassica napus*). Other crops, especially cabbages, are infected by this fungus. Field isolates can be ascribed consistently to one of two major pathotype groups, designated as aggressive and non-aggressive (Badawy and Hoppe, 1989) or virulent and avirulent (Gabrielson, 1983). Whereas non-aggressive isolates are restricted to the leaves and cause only minor losses, aggressive isolates give rise to systemic infections and lead finally to the symptoms of the blackleg syndrome in the hypocotyl region. During early infection stages, the fungus grows biotrophically in the intercellular space of the leaves. Later on, *L. maculans* penetrates the xylem vessels and spreads within the plant. At this stage, the fungus is latent and no obvious symptoms are detected. In the final stage of infection, the fungus reaches the hypocotyl region and switches its lifestyle to necrotrophy (Hammond *et al.*, 1985). Toxins of the sirodesmin family are synthesized in considerable amounts, and the massive growth of the fungus, its sexual and asexual sporulation, lead to stem canker formation, secondary rotting of the hypocotyl region and finally to the death of the plant (Hammond *et al.*, 1985; Hammond and Lewis, 1987).

The fast and reliable recognition of *L. maculans* and especially of the pathotype group is important for rapeseed cultivation. During the last few decades, the blackleg disease has gained increased importance due to the cultivation of the less resistant 00-cultivars. Spraying with fungicides could be limited or avoided if the pathotype group could be identified easily, because plants infected with non-aggressive strains would not have to be treated at all.

Aggressive and non-aggressive strains cannot be distinguished by their symptoms during the latent phase of infection. Morphological discrimination in pure culture is not applicable either, because of the high overall variation within the species. Several physiological approaches have been worked out in order to differentiate the pathotype groups. Classification by means of toxin production and pigment formation (Koch *et al.*, 1989) is possible, but these experiments are time consuming and not very reliable. Differentiation at the molecular level has also been achieved by electrophoretic karyotyping (Taylor *et al.*, 1991) and by means of RFLPs (Johnson and Lewis, 1990; Hassan *et al.*, 1991; Koch *et al.*, 1991). These approaches are reliable but rather tedious due to labelling of probes and the necessary blotting and hybridization steps.

We have developed a simple assay for the fast and reliable classification of the pathotype group of *L. maculans*, which is based on the RAPD-PCR technique (Williams *et al.*, 1990; Welsh *et al.*, 1991). DNA preparations of *L. maculans* strains were amplified with randomly designed nonamer and decamer oligonucleotide primers, which resulted in patterns of DNA fragments specific for each pathotype group. The method works with crude DNA samples, which can be prepared in less than a day, and the simple and fast random amplified polymorphic DNA (RAPD) technique including electrophoresis can be carried out in one additional day. The RAPD technique can be employed for the generation of pathotype-specific hybridization probes. After amplification and purification of several electrophoretically separated fragments, we were able to identify a PCR product that hybridizes specifically to the chromosomal DNA of aggressive strains. By means of this probe it is possible to detect infections with aggressive *L. maculans* strains even in infected leaves. There is no need for cultivation of the fungus, thus speeding up the analysis to less than two days between recognition of the first symptoms and the final diagnosis.

PROTOCOL

Materials

Hardware, chemicals, kits

1. Thermocycler : Biometra Trio-Thermoblock, Göttingen, Germany.
2. X-ray films: Kodak XAR-5.
3. DNA transfer membranes: Hybond N, Amersham.
4. Agarose : Seakem LE, FMC.
5. DIG DNA labelling kit: Boehringer Mannheim, Germany.
6. DIG luminescent detection kit: Boehringer Mannheim, Germany.

Primers

1. Primer 14: 5'-GCCGTCTACG-3'.
2. Primer 17: 5'-GGCATCGGCC-3'.

Primer synthesis was performed with an Applied Biosystems automatic synthesizer according to the manufacturer's instructions.

Enzymes

1. *Taq* polymerase and incubation buffer: Boehringer Mannheim, Germany.
2. *Hind*III: New England Biolabs.

Buffers, solutions

1. Extraction medium for DNA:
   1% SDS, 10 mM EDTA
2. TE buffer
   10 mM Tris-Cl, pH = 8.0
   0.1 mM EDTA
3. TBE buffer
   89 mM Tris-borate, pH = 8.0
   2 mM EDTA
4. TAE buffer
   40 mM Tris-acetate, pH = 8.0
   1 mM EDTA
5. Prehybridization buffer
   50% deionized formamide
   1 × Denhardt's solution (Denhardt, 1966)
   0.5 mg ml^{-1} salmon sperm DNA
   0.1% SDS
   5 × SSC
   50 mM sodium phosphate, pH = 7.0
6. Hybridization buffer
   50% deionized formamide
   1 × Denhardt's solution (Denhardt, 1966)
   0.1 mg ml^{-1} salmon sperm DNA
   0.1% SDS
   5 × SSC
   20 mM sodium phosphate, pH = 7.0
7. Washing solution 1
   2 × SSC
   0.1% SDS
8. Washing solution 2
   0.1 × SSC
   0.1% SDS
9. Chemiluminescence buffer 1
   0.1 M maleic acid, pH = 7.5
   0.15 M sodium chloride
10. Chemiluminescence buffer 2a
    1% Boehringer blocking reagent in buffer 1
11. Chemiluminescence buffer 2b
    0.5% blocking reagent in buffer 1
12. Chemiluminescence buffer 3
    0.1 M Tris-Cl, pH = 9.5
    0.1 M sodium chloride
    0.05 M $MgCl_2$

Methods

DNA preparation

1. Freeze a 7–10-day-old *L. maculans* culture, grown on V8-juice agar in a Petri dish, overnight and cover with liquid nitrogen. Scrape the frozen mycelium into a pre-cooled mortar and grind well with a pestle under liquid nitrogen.
2. Transfer the resulting fine powder into a 1.5 ml microcentrifuge tube and suspend in 1 ml extraction medium by vortexing.
3. Add 70 mg of solid sodium chloride, dissolve and incubate for 30 min on ice. Spin 15 min at 12,000 rpm in a microcentrifuge and transfer 1 ml of the supernatant into a fresh tube.
4. Add 100 mg of solid PEG 6000, dissolve, incubate for 1 h on ice, spin 15 min at 12,000 rpm and discard the supernatant.
5. Dissolve the pellet in 0.18 ml Tris-EDTA (TE) buffer, add 0.02 ml of 4 M LiCl, mix and precipitate the DNA with 0.2 ml isopropanol. Mix carefully and incubate on ice 30 min. Spin at 12,000 rpm for 15 min.
6. Discard the supernatant, wash the pellet once in 70% ethanol, dry briefly and dissolve in 0.05 ml TE buffer.
7. Check DNA concentration by analysis of 0.005 ml of each sample on an 0.8% agarose gel run in TBE buffer.
8. Dilute DNAs in TE buffer (normally 1:10).

RAPD-PCR and electrophoresis

1. Assay conditions for one sample (total volume: 50 μl):
   5 μl DNA (approximately 25 ng)
   5 μl 10 × incubation buffer
   4 μl dNTPs, 2.5 mM each
   4 μl primer (0.01 mg μl^{-1})
   3 μl 50 mM magnesium acetate
   28.5 μl H$_2$O
   0.5 μl *Taq* polymerase (2.5 units)
2. Mix the ingredients in a 0.5 ml microcentrifuge tube, spin down for a few seconds, and overlay with 0.05 ml of liquid paraffin.
3. Incubate in a thermocycler performing the following reaction profile:
   (i)   20 s at 95°C
   (ii)   60 s at 32°C
   (iii) 20 s at 72°C
   repeat (i) to (iii) 28 times
   (iv) 6 min at 72°C
   (v)   cool to 21°C
4. Transfer the lower phase into 1.5 ml microcentrifuge tubes.
5. Separate the amplified DNA fragments on a 1.2% agarose gel in TBE buffer, stain with ethidium bromide and visualize the results on a UV transilluminator.

Hybridization of genomic DNA with RAPD-derived probes

1. Digest 4 μg of restriction grade fungal DNA with 20 units of *Hind*III for 4 h at 37°C.

2. Separate restriction fragments on 0.8% agarose gels in TAE buffer and transfer to Hybond-N membranes by capillary transfer.
3. Incubate in prehybridization buffer for at least 5 h at 42°C in a waterbath shaker.
4. Label the probe (PCR product eluted from the gel) with digoxigenin according to the supplier's instructions.
5. Hybridize the filter with the denatured probe overnight in hybridization buffer at 42°C.
6. Rinse filter 4 times at room temperature in washing solution 1.
7. Wash the filter in a waterbath shaker according to the following regime (20 min per step):

   twice in washing solution 1 at 42°C
   twice in washing solution 1 at 55°C
   twice in washing solution 2 at 55°C.

Detection of hybridization patterns by chemiluminescence

The following steps are carried out at room temperature:

1. Incubate filter in buffer 1 for 5 min (100 ml per 100 cm^2).
2. Incubate filter in buffer 2a for 30 min (100 ml per 100 cm^2).
3. Incubate filter in buffer 2b containing 1/10,000 anti-dioxigenin-alkaline phosphatase (anti-DIG–AP) conjugate (20 ml per 100 cm^2).
4. Wash filter three times for 10 min in buffer 2b.
5. Wash filter three times for 10 min in buffer 1.
6. Equilibrate filter for 5 min in buffer 3.
7. Incubate filter in substrate solution for 5 min.

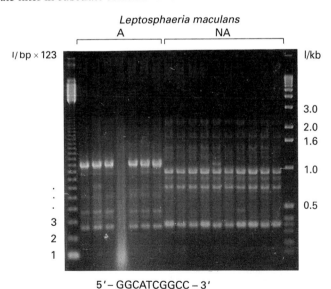

Fig 1.1. Differentiation between aggressive and non-aggressive strains of *L. maculans* using primer 17 (5'-GGCATCGGCC-3'). Left-hand markers: 123 bp-ladder; right-hand marker 1 kb-ladder; lanes 1–3 and 5–7: aggressive isolates; lanes 8–17: non-aggressive isolates.

Aggressive Non-aggressive

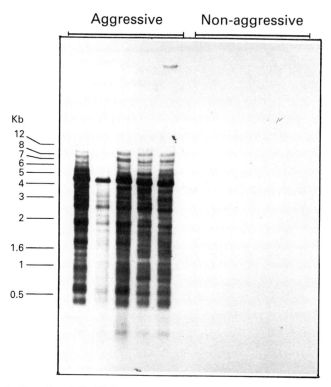

Fig 1.2. Southern blot of *Hind* III digested DNA from five aggressive (lanes 1–5, left) and five non-aggressive (lanes 6–10, right) *L. maculans* strains, hybridized with a PCR-derived band specific for aggressive stranins. Marker: 1 kb-ladder.

8. Allow filter to dry for 1–2 min on a Whatman filter; turn upside down once during this time.
9. Seal the damp filter in a plastic bag and incubate for 15 min at 37°C.
10. Expose filter to an X-ray film for a time between 2 min and 2 h, depending on the activity of the probe.

INTERPRETATION AND EXAMPLES OF RESULTS

Several of 20 tested nonamer and decamer random primers showed a clearcut differentiation between the two major pathotype groups after RAPD-PCR analysis. The best primer sequence for this purpose has the sequence 5'-GGCATCGGCC-3' (primer 17). An example is given in Fig. 1.1. With this specific primer, there is almost no variation between isolates within the two pathotype groups. Typical bands for the aggressive group have an electrophoretic mobility corresponding to 1100 and 340 bp, while diagnostic bands for non-aggressive strains have a size of 1000 and 350 bp. About 150 isolates of *L. maculans* were tested with the RAPD technique and all could be classified unequivocally with this primer.

DNA preparations can be very crude for RAPD analysis, and only very small amounts of DNA (roughly 25 ng) are required for one amplification reaction. It is, however, important that the DNA has a high molecular weight in order to avoid 'self-priming' with short fragments of denatured DNA, which compete with the RAPD primer resulting in a smear instead of distinct band patterns (Wöstemeyer *et al.*, 1991).

Amplification of DNA from aggressive *L. maculans* isolates with primer 14 (5'-GCCGTCTACG-3') gave a 600 bp fragment that could be shown to serve as a specific probe for a repetitive DNA element in aggressive strains. Figure 1.2 shows a Southern blot with *Hind*III digested genomic DNA of five aggressive and five non-aggressive isolates. The RAPD-PCR-derived, digoxigenin-labelled probe does hybridize to DNA from non-aggressive fungi. This probe (A-probe) can be used for tissue blot experiments or for dot blots of crude DNA preparations from infected rapeseed leaves. A different PCR product, the 350 bp fragment from non-aggressive strains amplified with primer 17 (Fig. 1.1) was shown to be a specific probe for DNAs from non-aggressive isolates (data not shown). This NA-probe can be used for similar purposes as the A-probe described above.

Compared with conventional PCR, RAPD-PCR has the advantage that no sequence information about the target DNA is required. Due to the short universal primer sequences, essentially any genomic DNA can be amplified. The technique is especially useful for comparing closely related genomes at the subspecies and pathotype level. Different species result in very divergent electrophoretic patterns. Thus, RAPD is suited to fill the analytical gap at the subspecies level where conventional methods often fail.

REFERENCES

Badawy, H. M. A. and Hoppe, H. H. (1989) Production of phytotoxic sirodesmins by aggressive strains of *Leptosphaeria maculans* differing in interactions with oil seed rape genotypes. *Journal of Phytopathology* 127, 146–157.

Denhardt, D. T. (1966) A membrane-filter technique for the detection of complementary DNA. *Biochemical and Biophysical Research Communications* 33, 641–646.

Gabrielson, R. L. (1983) Blackleg disease of crucifers caused by *Leptosphaeria maculans* (*Phoma lingam*) and its control. *Seed Science and Technology* 11, 749–780.

Hammond, K. E. and Lewis, B. G. (1987) Variation in stem infections caused by aggressive and non-aggressive isolates of *Leptosphaeria maculans* on *Brassica napus* var. *oleifera*. *Plant Pathology* 63, 53–65.

Hammond, K. E., Lewis, B. G. and Musa, T. M. (1985) A systemic pathway in the infection of oilseed rape plants by *Leptosphaeria maculans*. *Plant Pathology* 34, 53–65.

Hassan, A. K., Schulz, C., Sacristan, M. D. and Wöstemeyer, J. (1991) Biochemical and molecular tools for the differentiation of aggressive and non-aggressive isolates of the oilseed rape pathogen, *Phoma lingam*. *Journal of Phytopathology* 131, 120–136.

Johnson, R. D. and Lewis, B. G. (1990) DNA polymorphism in *Leptosphaeria maculans*. *Physiological and Molecular Plant Pathology* 37, 417–424.

Koch, E., Badawy, H. M. A. and Hoppe, H. H. (1989) Differences between aggressive and non-aggressive single spore lines of *Leptosphaeria maculans* in cultural characteristics and phytotoxin production. *Journal of Phytopathology* 124, 52–62.

Koch, E., Song, K., Osborn, T. C. and Williams, P. H. (1991) Relationship between pathogenicity and phylogeny based on restriction fragment length polymorphism in *Leptosphaeria maculans*. *Molecular Plant–Microbe Interactions* 4, 341–349.

Taylor, J. L., Borgmann, I. and Seguin-Swartz, G. (1991) Electrophoretic karyotyping of *Leptosphaeria maculans* differentiates highly virulent from weakly virulent isolates. *Current Genetics* 19, 273–277.

Welsh, J., Petersen, C. and McClelland, M. (1991) Polymorphisms generated by arbitrarily primed PCR in the mouse: application to strain identification and genetic mapping. *Nucleic Acids Research* 19, 303–306.

Williams, J. G. K., Kubelik, A. R., Livak, K. J., Rafalski, J. A. and Tingey, S. V. (1990) DNA polymorphisms amplified by arbitrary primers are useful as genetic markers. *Nucleic Acids Research* 18, 6531–6535.

Wöstemeyer, J., Schäfer, C., Kellner, M. and Weisfeld, M. (1991) DNA polymorphisms detected by random primer dependent PCR as a powerful tool for molecular diagnosis of plant pathogenic fungi. *Advances in Molecular Genetics* 5, 227–240.

2

A Monoclonal Antibody-based Immunoassay for the Detection of the Eyespot Pathogen of Cereals *Pseudocercosporella herpotrichoides*

F. M. Dewey[1] and R. A. Priestley[2]

[1] *Department of Plant Sciences, University of Oxford, South Parks Road, Oxford OX1 3RB, UK;*
[2] *present address, Delta Biotechnology, Castle Court, 59 Castle Boulevard, Nottingham NG7 1FD, UK.*

Background

Eyespot is an important disease of winter cereals in northern Europe, Canada, South Africa and New Zealand and is the cause of significant yield losses (Fitt, 1988; Polley and Thomas, 1991). The causal agent, *Pseudocercosporella herpotrichoides*, is one of a number of stem-base pathogens that attack cereals. Differentiation of the eyespot pathogen from the other stem-base pathogens, such as *Rhizoctonia cerealis*, the cause of sharp eyespot, *Microdochium nivale*, the snow mould pathogen and species of *Fusarium*, notably *F. avenaceum* and *F. culmorum*, is important. This is because *P. herpotrichoides* is the only stem-based pathogen which causes significant yield losses that respond to fungicide treatment (Fitt, 1988; Fitt *et al.*, 1988). However, specific identification of *P. herpotrichoides* is difficult because other stem-base pathogens may cause similar symptoms. At present, identification can only be confirmed in the laboratory by plating out infected, surface-sterilized material on appropriate media. The pathogen is slow growing and its presence is often masked by the other faster-growing stem-base pathogens or saprophytes. Thus, if the disease is suspected the farmer is generally advised to spray because laboratory confirmation of the disease may be too late for effective fungicide treatment. In order to reduce unnecessary applications of fungicides and thereby lower production costs and possible pollution, there is need for a rapid, simple assay such as an immunoassay that will confirm the presence of *P. herpotrichoides* quickly. Immunoassays for eyespot have been developed using antisera (polyclonal antibodies) raised against mycelial fragments, myce-lial extracts and culture fluids of *P. herpotrichoides*. Most of these antisera are not wholly satisfactory because they cross-react with one or more of the other stem-base pathogens. Bolik *et al.* (1987) reported that their antisera to *P. herpotrichoides* cross-reacted with *F. culmorum* and *Erysiphe graminis*. The antiserum raised by Lind (1988) against pathogen-specific, soluble proteins bands, was apparently more specific but it was apparently not tested against *R. cerealis*. Saunders and co-workers (see Smith *et al.*, 1990; Cagnieul and Lefebvre, 1991) raised antiserum to freeze-dried culture fluids of *P. herpotrichoides* which did not cross-react with *R. cerealis* or *Fusarium* species and has subsequently been shown not to react with *M. nivale* (DuPont, personal communication). More recently, Poupard *et al.* (1991) reported raising antisera to a protein from *P. herpotrichoides*, purified by electrophoresis, which does not recognize *R. cerealis* or *M. nivale* or any other stem-base fungi. However, no details have been given about how the protein was derived or its relative molecular weight (M_r). We have raised antibodies, both polyclonal and mono-clonal, to a water-soluble protein/glycoprotein fraction from *P. herpotrichoides* and used these to develop a double antibody sandwich immunoassay.

PRODUCTION OF *PSEUDOCERCOSPORELLA*-SPECIFIC ANTISERUM AND MONOCLONAL ANTIBODIES

Antigens for immunization were derived from a phosphate buffered saline(PBS) extract of freeze-dried mycelium ($20\,mg\,ml^{-1}$) of a wheat pathotype, MBC-sensitive isolate, of *P. herpotrichoides* grown in liquid culture on potato dextrose broth at 17°C for 10 days, from which the carbohydrates and high molecular weight (>300 kD) glycoproteins and proteins had been removed by fractional precipitation with ammonium sulphate (for

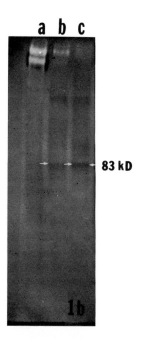

Fig 2.1. Western blot of 7.5% native polyacrylamide gel of PBS extracts from freeze dried mycelium of *Pseudocercosporella herpotrichoides* probed with *Pseudocercosporella* monoclonal antibody PH-10. Lanes a and c ammonium sulphate precipitates of PBS extracts from mycelium: (a) high molecular weight material precipitated with 30% ammonium sulphate; (c) lower molecular weight material precipitated by 70% ammonium sulphate but not 30%; (b) PBS extract from whole mycelium. Note immunopositive band (arrow) in lanes b and c at 83 kD.

details see Priestley and Dewey, 1993). The antigen fraction contained a protein with an approx M_r 83 kD (Fig. 2.1, see also Dewey *et al.*, 1991).

Balb/C mice were given six intraperitoneal injections at approx. two week intervals. Each injection contained approx. 2 mg of the immunogen (approx. 31 μg of protein). New Zealand white rabbits were given five intramuscular injections each containing a total of 22 mg of the immunogen (approx 600 μg of protein) in PBS, emulsified for the first injection with Freund's complete adjuvant and with Freund's incomplete adjuvant for subsequent injections. The antisera raised in rabbits had a higher titre than that raised in mice and was considerably more specific (Table 2.1). Splenocytes from mice immunized as described above were used to raise hybridomas. Hybridoma supernatants were screened by ELISA using wells coated initially with PBS-soluble extracts from lyophilized mycelium of *P. herpotrichoides* and subsequently against surface washings of cultures of other stem-base fungi and extracts from infected and healthy wheat plants. From two fusions relatively few hybridoma cell lines were obtained and of these, only one cell line (PH 10) secreted monoclonal antibodies (MAbs) that were sufficiently specific for the development of a detection assay (Figs 2.2 and 2.3). MAb-PH10 recognized all of the 15 isolates of the pathogen tested and showed no significant cross-reactivity with the other

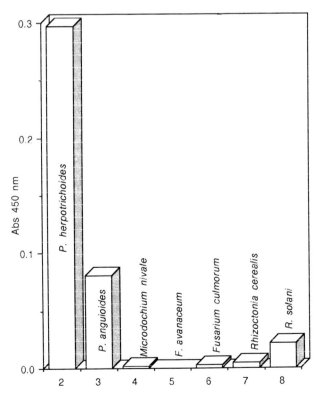

Fig 2.2. Absorbance readings from ELISA tests of antisera raised against an ammonium sulphate precipitate fraction from mycelial extracts of *Pseudocercosporella herpotrichoides* tested against wells coated with PBS extracts of freeze-dried mycelia (2 mg ml⁻¹) of *P. herpotrichoides, Microdochium nivale* and *Rhizoctonia solani.*

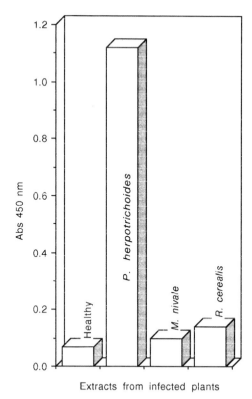

Fig 2.3. Absorbance values from DAS-ELISA tests on extracts from healthy plants and plants infected with *Pseudocercosporella herpotrichoides*, *Microdochium nivale* and *Rhizoctonia cerealis*.

stem-base fungi. Heat and periodate treatment of the antigen indicate that the MAb-PH10 recognizes a protein epitope on a glycoprotein (see Priestley and Dewey, 1993 for details).

Table 2.1. Absorbance readings from ELISA tests of antisera raised against an ammonium sulphate precipitate fraction from mycelial extracts of *Pseudocercosporella herpotrichoides* tested against wells coated with PBS extracts of freeze-dried mycelia (2 mg m^{-1}) of *P. herpotrichoides*, *Microdochium nivale* and *Rhizoctonia solani*.

	Reciprocal dilutions of antiserum				
	6400	12,800	25,600	51,200	102,400
Rabbit PH-10 antiserum					
P. herpotrichoides	1.22	1.07	0.85	0.61	0.37
M. nivale	0.53	0.34	0.21	0.13	0.10
Mouse PH-10 antiserum					
P. herpotrichoides	1.14	0.93	0.66	0.52	0.34
M. nivale	0.78	0.58	0.31	0.20	0.13
R. solani	0.60	0.40	0.24	0.16	0.12

DOUBLE ANTIBODY SANDWICH ENZYME LINKED IMMUNOSORBENT ASSAY (DAS-ELISA)

Materials

DYNATECH IMMULON 2 Microtitre Immunoassay plates.
Acetate buffer pH 4.8.
PBS – phosphate buffered saline pH 7.2.
PBST – PBS with 0.05% Tween 20.
Monoclonal antibody – MAb-PH10.
Rabbit polyclonal antibody (RPAb-PH10).
Goat anti-rabbit IgG (whole molecule) peroxidase conjugate (Sigma A-4914).
Substrate buffer – made up fresh each day from stock solutions of 0.2 M sodium acetate and 0.2 M citric acid.
Substrate stock – 10 mg TMB per ml undiluted DMSO, stored at 4°C.
Working substrate – 10 ml substrate buffer + 100 μl stock substrate solution + 3 μl 30% H_2O_2.

Methods

1. Coat microtitre wells with MAb-PH10, 50 μl per well.
2. Wash 4 × 5 min with PBST, invert and flick out remaining liquid, dry and store at 4°C or use directly.
3. Block pre-coated wells for 10 min with 1% BSA in PBS.
4. Wash 4 × 5 min with PBST, invert and flick out remaining liquid.
5. Add 50 μl of sample extract to each well, samples are prepared as follows:
   (a) Place weighed sample (either 3.5 cm of stem base of seedlings or individual leaf sheaths of older plants) in a plastic bag containing water-proof silicon carbide paper (660A).
   (b) Add appropriate amount of acetate buffer (1/5 w/v).
   (c) Crush sample in bag using BIOREBA hand-held homogenizer.
   (d) Remove liquid from sample bag by squeezing the sample and collecting the liquid in a free corner.
   (e) Place supernatant in an Eppendorf and spin in a microfuge for 2 min full speed.
   (f) Plate out supernatant in triplicate on to blocked plates, 50 μl per well.
6. Wash plate 4 × 5 min with PBST as in step 2.
7. Place 50 μl of rabbit antiserum, freshly diluted, 1 in 500, into 0.1 M acetate buffer pH 4.8, in each well, incubate at room temperature (RT) for 1 h.
8. Wash plate five times with PBST as in step 2.
9. Add 50 μl of anti-rabbit peroxidase conjugate (Sigma A 4914), freshly diluted, 1 in 5000 into PBST, to each well, incubate at RT for 1 h.
10. Wash plate 4 × 5 min with PBST.
11. Add 50 μl of freshly prepared working substrate to each well, incubate for 5 min.
12. Stop the reaction by adding 3 M H_2SO_4 to each well and read absorbance at 450 nm on ELISA plate reader using the healthy material as a control blank.

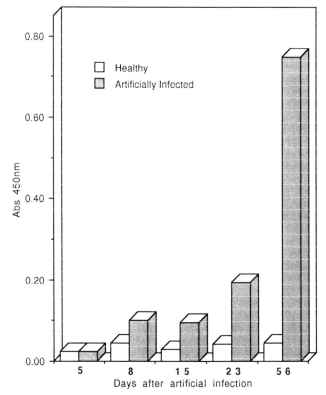

Fig 2.4.   Absorbance values from DAS-ELISA tests of extracts from stem bases of healthy and artificially infected wheat plants 5, 8, 15, 23 and 56 days after infection. The infected plants were each sprayed at 5 days old with approximately 1 ml of a spore suspension of *Pseudocercosporella herpotrichoides* containing 5×10^5 spores. Note: signs of eyespot lesions were not visible on plants before 21 days after inoculation.

RESULTS

The above assay system has proved to be effective, specific and sensitive. Routinely low absorbance values have been obtained with extracts from healthy tissues and tissues infected with other stem-base fungi (Fig. 2.2), and consistently high readings have been obtained with tissues infected with *P. herpotrichoides*. In tests with young, artificially inoculated plants, using extracts from infected and healthy stem bases or leaf sheaths, the pathogen has been detected before symptoms are apparent, i.e. 8 to 15 days after inoculation (Fig. 2.4).

LIMITATIONS

In developing the assay, we encountered considerable cross-reactivity with extracts from healthy plant material using antigen-coated wells. These difficulties were overcome by developing a double antibody sandwich ELISA (DAS-ELISA) in which the monoclonal antibody PH-10 was used as a capture antibody and the rabbit antiserum as the detector antibody. The reverse procedure, with wells coated with the rabbit polyclonal antiserum, gave only background absorbance values, which may indicate that only a single epitope or a small number of closely positioned epitopes on the antigen are recognized by the MAb-PH10. We found that low pH and the presence of Tween 20 in the extraction buffer is important in keeping background values to a minimum. The use of waterproof silicon carbide paper helps to increase the amount of antigen extracted. Plant tissues less than 12 days old were found to contain a significant amount of wheat germ agglutinin (WGA, a lectin) which bound to the rabbit antibodies. We found that the problem could be circumvented by using an acetate buffer at pH 4.8. At this pH the WGA dimer dissociates and its binding abilities are therefore reduced.

Preliminary studies indicate that the assay can also be used for quantitative measurements of the pathogen within infected tissue.

ACKNOWLEDGEMENT

The authors acknowledge DowElanco for financial support and provision of material infected with *P. herpotrichoides*.

REFERENCES

Bolik, M., Casper, R. and Lind, V. (1987) Einsatz serologischer und gelektrophoretischer Verfahren zum Nachweis von *Pseudocercosporella herpotrichoides*. *Zeitschrift für Pflanzenkrankheiten und Pflanzenschutz* 94, 449–456.

Cagnieul, P. and Lefebvre, A. D. (1991) Diagnostic kit (diagnolab) for cereal eyespot. *Proceedings of the Third International Conference on Plant Diseases, Bordeaux*, pp. 547–553.

Dewey, F. M., Priestley, R. A. and Sturz, A. V. (1991) Antigenic material, antibodies raised thereto, hybridomas for producing antibodies and immunoassays incorporating such antibodies. In: *European Patent Number 91307797.0.*, taken out by DowElanco.

Fitt, B. D. L. (1988) *Eyespot Disease of Cereals*. Home Grown Cereals Authority, Hamlyn House, Highgate Hill, London, N19 5PR, 52 pp.

Fitt, B. D. L., Goulds, A. and Polley, R. W. (1988) Eyespot (*Pseudocercosporella herpotrichoides*) epidemiology in relation to prediction of disease and severity and yield loss winter-wheat – a review. *Plant Pathology* 37, 311–329.

Lind, V. (1988) Estimation of the resistance to *Pseudocercosporella herpotrichoides* by a serological method. In: *Proceedings of the Conference of the Cereal Section of Eucarpia, Wageningen, Netherlands*, Pudoc, Wageningen, pp. 145–149.

Polley, R. W. and Thomas, M. R. (1991) Surveys of diseases of winter wheat in England and Wales 1976–1988. *Annals of Applied Biology* 119, 1–20.

Poupard, P., Grare, S. and Cavelier, N. (1991) Using ELISA method for assessment of *in vitro* development of strains of *P. herpotrichoides*, the cereal eyespot fungus. *Proceedings of the Third International Conference on Plant Diseases, Bordeaux*, pp. 655–662.

Priestley, R. A. and Dewey, F. M. (1993) Development of a monoclonal antibody immunoassay for the eyespot pathogen *Pseudocercosporella herpotrichoides*. *Plant Pathology* 42, 403–412.

Smith, C. M., Saunders, D. W., Allison, D. A., Johnson, L. E. B., Labit, B., Kensall, S. J. and Hollomon, D. W. (1990) Immunodiagnostic assay for cereal eyespot: novel technology for disease detection. *Proceedings of the Brighton Crop Protection Conference – Pests and Diseases*, pp. 763–770.

3

DNA Probe for R-type of Eyespot Disease of Cereals
Pseudocercosporella herpotrichoides

P. Nicholson and H. N. Rezanoor

Cambridge Laboratory, John Innes Centre for Plant Science Research, Colney Lane, Norwich NR4 7UJ, UK.

Background

The fungus *Pseudocercosporella herpotrichoides* (Fron.) Deighton [Teleomorph: *Tapesia yallundae* Wallwork and Spooner] causes eyespot, an economically important disease of cereals in the UK, other European countries and other temperate cereal growing regions (Ponchet, 1959; Breuhl *et al.*, 1968; Banyer and Kuiper, 1976). The fungus infects the stem base where it interferes with translocation of nutrients and weakens the stem. Severely infected plants may collapse making harvesting difficult. Increased humidity in the lodged crop leads to infection of the ears and further loss of yield and quality. Two main pathotypes, W and R, are widespread in the UK. W-types are highly pathogenic on wheat but are only slightly pathogenic on rye and certain grasses whereas R-types are generally equally pathogenic on wheat and rye (Lange-de la Camp, 1966; Scott and Hollins, 1980). When cultured *in vitro*, R- and W-types generally differ in morphology and colour (Hollins *et al.*, 1985; Creighton, 1989). However, some isolates cannot be identified without recourse to pathogenicity testing (Hollins *et al.*, 1985; Gallimore *et al.*, 1987). Pathogenicity testing of an isolate is labour intensive, generally taking 12–16 weeks, and relies upon observing the depth of fungal penetration through the leaf sheaths around the stem base.

Differences in infection processes and polymorphisms in isozymes and restriction fragment length polymorphisms (RFLPs) of ribosomal DNA (rDNA) and mitochondrial DNA (mtDNA) suggest that the R-types may be distinct from W-types (Julian and Lucas, 1990; Daniels *et al.*, 1991; Nicholson *et al.*, 1991, 1993; Priestley *et al.*, 1992).

Polyclonal and monoclonal antibody-based immunoassays have been developed for the detection of *P. herpotrichoides* (Dewey, 1988; Smith *et al.*, 1990) but apparently neither can differentiate the W- and R-pathotypes. In addition, such antibodies tend to crossreact with a related, though less damaging fungus *P. anguioides* which is also often present on the stem base.

We have identified a DNA marker which clearly differentiates R-type and W-type isolates by only hybridizing to R-type isolates and has been used to detect the fungus in inoculated plant material. Quantification of the fungus in infected tissue using DNA-based techniques may offer a novel approach for evaluating the resistance of hosts and/or the pathogenicity of isolates.

DEVELOPMENT OF R-TYPE SPECIFIC PROBE

Mycelium was scraped from the surface of 14-day-old colonies of an R-type isolate (C79/243/1) and W-type isolate (C87/631/1) on potato dextrose agar (PDA) and used to inoculate 50 ml of yeast extract glucose (YEG) liquid medium (Nicholson et al., 1991). Cultures were grown on an orbital shaker at 25°C in darkness for 10 days and mycelium was harvested by filtration on to Whatman no. 1 filter paper discs. The mycelium was frozen in liquid nitrogen and ground to a fine powder with a mortar and pestle and DNA was extracted and purified by the method of O'Dell et al. (1989).

Total genomic DNA ($3 \mu g$) of the two isolates was restricted separately with EcoRI and size fractionated by electrophoresis through agarose gel (0.8%). DNA in the size range 2–8 kb was electrophoresed on to DEAE paper and eluted into Eppendorf tubes as described in Sambrook et al. (1989). The size-fractionated DNA fragments were ligated into pUC18 cleaved with EcoRI and used to transform competent E. coli DH5α cells (Sambrook et al., 1989). Recombinant clones were selected on L agar plates containing ampicillin 50 $\mu g \, ml^{-1}$, Xgal (5-bromo-4-chloro-3-indolyl-β-D-galactopyranoside) 20 mg ml^{-1} and IPTG (isopropyl-β-D-thiogalactopyranoside) 24 mg ml^{-1}. Recombinant colonies were transferred to replicated ampicillin plates and immobilized on nitrocellulose filters as described by Grunstein and Hogness (1975).

Total genomic DNA (25 ng) of isolate C79/243/1 (R-type) and isolate C87/631/1 (W-type) was mechanically sheared in a sonic bath prior to labelling with ^{32}P dCTP (Amersham International) to high specific activity by the 'oligo-labelling' method (Feinberg and Vogelstein, 1984). The recombinant DNA clones on nitrocellulose filters were hybridized sequentially with ^{32}P-labelled total genomic DNA of the R- and W-type isolate to identify clones which hybridized only to DNA of either, but not both, pathotypes.

Putative pathotype-specific clones were grown in small-scale cultures and plasmid DNA was isolated according to Sambrook et al. (1989). A single clone, R04, from the R-type isolate produced a strong signal when hybridized to R-type DNA and a negligible signal with W-type DNA. Plasmid DNA of clone R04 was labelled as described above and hybridized on to size fractionated EcoRI digested total genomic DNA of a wide range of isolates of P. herpotrichoides, P. anguioides and P. aestiva immobilized on nylon membranes (Hybond N+, Amersham International). All membranes were prehybridized, hybridized and washed as described in Nicholson et al. (1993).

DNA HYBRIDIZATION PROTOCOL

Materials

1. Extraction buffer: 100 mM Tris HCl pH 8.0, 100 mM NaCl, 50 mM EDTA, 20 gl^{-1}SDS.
2. Proteinase K (Sigma Chem Co.) 0.1 mg ml^{-1}.

3. Denhardt's solution: 20 g l^{-1} gelatine, 20 g l^{-1} ficoll 400, 20 g l-1 polyvinyl pyrroli-done 360, 100 g l^{-1} SDS, 50 g l^{-1} sodium pyrophosphate.
4. 5 × HSB 0.6 mM sodium chloride, 10 mM Pipes pH 6.8, 1 mM EDTA.
5. Salmon sperm DNA (100 μg ml^{-1}), freshly boiled in a microwave.
6. Buffer A: mixture of H$_2$O: 5 × HSB: Denhardt's solution: salmon sperm DNA (60 : 20 : 10 : 10).
7. 20 × SSC: 3 M sodium chloride, 0.3 M sodium citrate, pH 7.0.
8. DNA of clone R04 (P. Nicholson, Cambridge Laboratory, John Innes Centre, Norwich NR4 7UJ, UK).

Method for the detection P. herpotrichoides *in plant tissue*

1. Sections of 2 cm of each of five seedling stem bases, including all of any lesion present, are removed, weighed and freeze-dried or ground in liquid nitrogen.
2. The powder is added to a tube containing 5 ml of extraction buffer and 40 μl of Proteinase K stock and incubated for 1–2 h at 65°C with occasional mixing.
3. After extraction with an equal volume of phenol/chloroform saturated with 10 mM Tris HCl pH 8.0, the mixture is centrifuged at 3000g for 15 min.
4. The aqueous phase is removed and added to an equal volume of chloroform, mixed and centrifuged as above.
5. The aqueous phase is mixed with 2 volumes of ethanol and precipitated at −20°C overnight.
6. DNA is pelleted by centrifugation as above, the liquid removed and the pellet dried.
7. DNA is dissolved in 10 mM Tris HCl and 1 mM EDTA pH 8.0 equivalent to 2.5 mg of fresh weight per ml.
8. For slot blots (rye cultivars only), DNA is diluted to 0.5 mg fresh weight equivalent and a fourfold dilution series prepared. Samples are transferred to nylon membranes (Hybond N+) using a slot-blotting apparatus (Bio-Rad) according to the manufacturer's instructions.
9. DNA from 5 mg fresh weight equivalent is restricted with *Eco*RI, electrophoresed through an 0.8% agarose gel using standard procedures (Sambrook *et al.*, 1989) and transferred to nylon membranes (Hybond N+) by alkaline Southern blotting (Reed and Mann, 1985).
10. All membranes are prehybridized in buffer A at 65°C for 16 h.
11. Probe R04 DNA is labelled with ^{32}P-dCTP by single-stranded DNA synthesis using random oligonucleotides as primers (Feinberg and Vogelstein, 1984).
12. Membranes are hybridized to ^{32}P-labelled R04 probe in buffer A at 65°C for 16 h.
13. Membranes are washed at 65°C (2 × 15 min) with 2 × SSC, 0.1% SDS and then at the same temperature (2 × 15 min) with 0.2% SSC, 0.1% SDS.
14. Membranes are blotted on to filter paper, wrapped in plastic film (Saran wrap) and exposed to photographic film at −70°C between two intensifying screens for 1–5 days.

RESULTS AND LIMITATIONS

This clone hybridized to a single *Eco*RI fragment in all isolates tested to date known to be pathogenic to rye. Three fragment sizes (11.6, 6.7 and 4.5 kb) have been observed among the R-type isolates tested. Fragment size does not appear to be related to the origin of the isolate.

An extremely weak signal was detected in DNA of W-type isolates following prolonged exposure but was not detectable following normal exposure (Fig. 3.1). In addition, this clone does not hybridize to isolates of *P. anguioides* under the conditions described above. DNA extracted from rye seedlings infected with an R-type isolate hybridized in slot blots to R04 whereas uninoculated seedlings or seedlings inoculated with W-type isolates did not. Co-purified substances in the extractions interfered with the binding of DNA resulting in background haloes in the higher DNA concentrations equivalent to, or greater than, 0.125 mg of fresh weight material. In preliminary tests the probe has clearly detected 20 pg of purified fungal DNA in a dilution series run alongside R-type infected material.

Similar slot blots prepared from wheat seedlings have not revealed differences in the levels of hybridization because all samples, including those from uninoculated plants hybridized equally to probe R04 (results not shown).

To circumvent this problem, DNA from seedlings of wheat varieties is digested with *Eco*RI before electrophoresis and transfer to nylon membranes by Southern blotting for hybridization with probe R04. The gel loadings are made according to DNA quantities representing similar fresh weights of plant material in order to quantify infection. In certain lines and varieties (e.g. VPM1), a hybridizing fragment may be observed in all samples, including uninoculated control plants, which may, at least in part, be the cause of the background signal detected in slot blots. The presence of the R-type can still be observed where such additional fragments appear if the sizes of the fragments are not identical to that for the R-type. This system has been used to detect *P. herpotrichoides* (R-type) in resistant and susceptible wheat lines. Hybridization to clone R04 was

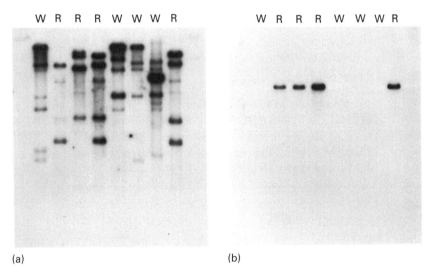

(a)   (b)

Fig 3.1. (a) Autoradiograph of total genomic DNA of eight isolates of *Pseudocercosporella herpotrichoides* digested with *Eco*RI and probed with M13 to produce genetic fingerprints of the isolates. (b) Autoradiograph of the same eight isolates as in (a) probed with R-type-specific probe (R04).

considerably greater in susceptible hosts than in resistant ones suggesting significant differences in the levels of active fungus and host infection.

REFERENCES

Banyer, R. J. and Kuiper, J. (1976) Aerial survey of eyespot lodging of wheat in southern New South Wales. *Australian Plant Pathology Society Newsletter* 5, 25–28.

Breuhl, G. W., Nelson, W. L., Koehler, F. and Vogel, O. A. (1968) Experiments with *Cercosporella* foot rot (straw breaker) disease of winter wheat. *Washington Agricultural Experiment Station Bulletin* No. 694, 14pp.

Creighton, N. F. (1989) Identification of W-type and R-type isolates of *Pseudocercosporella herpotrichoides*. *Plant Pathology* 38, 484–493.

Daniels, A., Lucas, J. A. and Peberdy, J. F. (1991) Morphology and ultrastructure of W and R pathotypes of *Pseudocercosporella herpotrichoides* on wheat seedlings. *Mycological Research* 95, 385–397.

Dewey, F. M. (1988) Development of immunological diagnostic assays for fungal plant pathogens *Brighton Crop Protection Conference – Pests and Diseases 1988*, pp. 777–786.

Feinberg, A. P. and Vogelstein, B. (1984) Addendum: a technique for radiolabelling DNA restriction endonuclease fragments to high specific activity. *Analytical Biochemistry* 137, 266–267.

Gallimore, K., Knights, I. K. and Barnes, G. (1987). Sensitivity of *Pseudocercosporella herpotrichoides* to the fungicide prochloraz. *Plant Pathology* 36, 290–296.

Grunstein, M. and Hogness, D. S. (1975) Colony hybridisation. A method for the isolation of cloned DNAs that contain a specific gene. *Proceedings of the National Academy of Sciences of the United States of America* 72, 3961–3965.

Hollins, T. W., Scott, P. R. and Paine, J. R. (1985) Morphology, benomyl resistance and pathogenicity to wheat and rye of isolates of *Pseudocercosporella herpotrichoides*. *Plant Pathology* 34, 369–379.

Julian, A. M. and Lucas, J. A. (1990) Isozyme polymorphism in pathotypes of *Pseudocercosporella herpotrichoides* and related species from cereals. *Plant Pathology* 39, 178–190.

Lange-de la Camp, M. (1966) Die Wirkungsweise von *Cercosporella herpotrichoides* Fron., dem erreger der Halmbruchkrankheit des Getreides. II. Agressivitat des Erregers. *Phytopathologische Zeitschrift* 56, 155–190.

Nicholson, P., Hollins, T. W., Rezanoor, H. N. and Anamthawat-Jonsson, K. (1991) A comparison of cultural, morphological and DNA markers for the classification of *Pseudocercosporella herpotrichoides*. *Plant Pathology* 40, 584–594.

Nicholson, P., Rezanoor, H. N. and Hollins, T. W. (1993) Classification of a world-wide collection of isolates of *Pseudocercosporella herpotrichoides* by RFLP analysis of mitochondrial and ribosomal DNA and host range. *Plant Pathology* 42, 58–66.

O'Dell, M., Wolfe, M. S. Flavell, R. B., Simpson, C. G. and Summers, R. W. (1989) Molecular variation in populations of *Erysiphe graminis* on barley, oats and rye. *Plant Pathology* 38, 340–351.

Ponchet J. (1959) La maladie du pietin-verse des cereales: *Cercosporella herpotrichoides* Fron. Importance agronomique, biologie, epiphytologie. *Annales des Epiphyties* 10, 45–98.

Priestley, R. A., Dewey, F. M., Nicholson, P. and Rezanoor, H. N. (1992) Comparison of isoenzyme and DNA markers for differentiating W-, R- and C-pathotypes of *Pseudocercosporella herpotrichoides*. *Plant Pathology* 41, 591–599.

Reed, K. C. and Mann, D. A. (1985) Rapid transfer of DNA from agarose gels to nylon membranes. *Nucleic Acids Research* 13, 7207–7221.

Sambrook, J., Fritch, E. F. and Maniatis, T. (1989) *Molecular Cloning: A Laboratory Manual*, 2nd edn. Cold Spring Harbor Press.

Scott P. R. and Hollins T. W. (1980) Pathogenic variation in *Pseudocercosporella herpotrichoides. Annals of Applied Biology* 78, 269–279.

Smith, C. M., Saunders, D. W., Allison, D. A., Johnson, L. E. B., Labit, B., Kendall, S. J. and Hollomon, D. W. (1990) Immunodiagnostic assay for cereal eyespot: novel technology for disease detection. *Brighton Crop Protection Conference – Pests and Diseases 1990*, pp. 763–770.

4

Detection of *Spongospora subterranea* by ELISA Using Monoclonal Antibodies

J. G. HARRISON, R. LOWE, A. WALLACE AND
N. A. WILLIAMS

Scottish Crop Research Institute, Invergowrie, Dundee DD2 5DA, UK.

INTRODUCTION

Powdery scab, caused by infection of potato by the fungus *Spongospora subterranea* leads to unsightly blemishes on tubers that reduce the market value of the crop. The yield of tubers can also be reduced if the disease is severe. In addition, *S. subterranea* is the vector of potato mop-top virus (Jones and Harrison, 1969) which can cause serious damage to some cultivars. The pathogen can survive for many years in field soil as resting spores produced in large numbers in tuber scabs. Resting spores are an important source of inoculum even with long crop rotations. The disease can also be introduced into a crop by planting seed tubers contaminated with resting spores. Disease incidence has increased recently in the UK, possibly because particularly susceptible cultivars are being grown and because the use of irrigation has become widespread. Wet soil conditions favour the disease (Wale, 1987). No cultivar is immune to powdery scab and the only effective form of control is to plant clean seed tubers into uncontaminated land. If completely pathogen-free land is unavailable a knowledge of numbers of resting spores present could allow the risk of disease to be assessed, taking into account the drainage characteristics of the land and the disease susceptibility of the potato variety to be grown.

Harrison *et al.* (1993) have developed an ELISA system, using polyclonal antiserum to resting spores, for the detection of tuber contamination. Resting spores in soil, however, were only weakly detected even when unrealistically high numbers were present. Soil contamination is currently assessed using bait plants (S. J. Wale, personal communication). The roots of young, intact tomato plants, which are also readily infected by *S. subterranea*, are placed in water to which soil is added. Resting spores germinate by releasing zoospores that infect the root hairs, where the plasmodial stage of the fungus is formed. Eventually large numbers of secondary zoospores are released from the roots and infect newly-formed root hairs, again producing plasmodia. Numbers of plasmodia are estimated by microscopical inspection, a laborious process. A sensitive ELISA system

23

for the quantification of plasmodia in the roots of bait plants, or for estimating numbers of zoospores released from resting spores would hasten soil testing.

S. subterranea is an obligate pathogen and therefore cannot be cultured outside the host. The zoospore stage is the only free-living form of the pathogen. All other forms are intimately associated with host cells from which it would be difficult to separate the fungus. Zoospores were used as the antigen for monoclonal antibody (MAb) production in the work described here to maximize the likelihood of obtaining hybridoma cell lines secreting antibodies recognizing only *S. subterranea* and not also the host.

PROTOCOL

Materials

ELISA   Nunc-Immuno Maxisorp F96 multi-well plates (Life Technologies).
Bovine serum albumin (BSA) (Product no. A-7030; Sigma Chemical Co.).
Anti-mouse IgG alkaline phosphatase conjugate (Product no. A-1047; Sigma Chemical Co.).
Anti-mouse IgG peroxidase conjugate (Product no. A–3673; Sigma Chemical Co.).
P-nitrophenyl phosphate tablets (Product no. N–2765; Sigma Chemical Co.).
3,3',5,5'-tetramethylbenzidine dihydrochloride tablets (Product no. T–3405; Sigma Chemical Co.).
Phosphate–citrate buffer with sodium perborate, capsules (Product no. P–4922; Sigma Chemical Co.).

Methods

Preparation of zoospore suspensions   Potato tubers with powdery scab lesions from a local crop were washed and air-dried. Cystosori (spore balls of *S. subterranea* each consisting of a mass of resting spores) inevitably with some tuber debris and soil, were scraped from lesions, air-dried, large particles removed by sieving, and stored at 4°C. Aliquots of 100 mg of cystosori were each mixed with 100 ml distilled water and incubated at 12–14°C in darkness for 4–5 days to encourage the release of zoospores. The resulting suspensions were filtered through Whatman no. 540 paper to remove cystosori, soil particles and tuber debris. Zoospore concentrations were determined using a haemocytometer. Either (a) PBS was added to the zoospore suspensions to give a concentration of 0.01 M and zoospore numbers adjusted to 4×10^6 ml^{-1} for immunization, or (b) 293 mg NaHCO$_3$, and 159 mg N$_2$CO$_3$ were added per 100 ml suspension (0.05 carbonate buffer, pH 9.8) and zoospore numbers adjusted to 2×10^5 ml^{-1}, for coating microtitre plates for screening for antibodies by plate-trapped antigen ELISA (PTA-ELISA). Suspensions of zoospores which had not been fixed were stored at −20°C in 1 ml aliquots for immunization and in 20 ml aliquots for use in ELISAs.

Production of MAbs   Five 10-week-old mice were given intraperitoneal injections of 1×10^6 zoospores in 0.25 ml PBS, estimated to contain *ca.* 0.3 μg protein, at 3 week intervals and at 7 and 3 days before fusion. The mice, designated nos 1–5, were given totals of 4, 9, 9, 10 and 12 injections respectively. Hybridoma cell lines producing antibodies that recognized zoospores but not comminuted healthy tomato roots were expanded *in vitro*. Cell supernatant fluids were concentrated and partially purified by ion-exchange chromatography and used at a dilution of 1 : 50 with TBS.

ELISA Cell supernatants were screened for the presence of antibodies recognizing zoospores of *S. subterranea* by indirect PTA-ELISA, following a technique similar to that described by Mohan (1988). Zoospores suspended in carbonate buffer were incubated overnight at 4°C in wells of microtitre plates (coating wells) followed by successive incubations for 60 min at 37°C with 1% bovine serum albumin (BSA) in Tris buffered saline (TBS), cell supernatant fluid and antimouse IgG peroxidase conjugate: 100 μl of each were placed in each well and wells were washed three times with TBS with Tween (TBST) after each incubation. Finally 100 μl of tetramethylbenzidine hyrochloride (enzyme substrate) in phosphate–citrate buffer were added to each well and the plates incubated at 37°C. The reaction was stopped by the addition of 50 μl 2 N H_2SO_4 to each well and absorbance at 450 nm (A450) was measured. Cell supernatant fluid was substituted with blood serum from an inoculated mouse, diluted 1 : 1000 with TBS, in one well per plate as a positive control. Alternatively, and in all ELISAs that included host plant material, antibody was visualized with anti-mouse IgG alkaline phosphatase conjugate and *p*-nitrophenyl phosphate (enzyme substrate) in diethanolamine–HCl buffer, pH 9.8, and absorbance determined at 405 nm (A405) several times during incubation at 37°C. Cell supernatants were also tested for antibodies recognizing uninfected tomato roots, plasmodia-infected tomato roots and cystosori of *S. subterranea* by ELISA. Wells were coated with suspensions in carbonate buffer of healthy roots, plasmodia-infected roots (100 mg fresh weight of root comminuted with a glass tissue grinder in 10 ml) or cystosori (1 × 10^4 ml^{-1}).

Reaction with other microorganisms Recognition by the MAbs of 31 mostly soil-inhabiting fungi other than *S. subterranea* and of *Erwinia carotovora*, a bacterium that is commonly found on potato tubers, was determined by indirect PTA-ELISA with peroxidase conjugate. The organisms were grown in Elliott's liquid medium (agar omitted) (Elliott *et al.*, 1966), removed from the culture solutions by Buchner filtration (fungi) or by centrifugation (*Erwinia*) and comminuted in carbonate buffer. The concentration of each was adjusted to an estimated equivalent (fresh weight: volume) of 2 × 10^5 zoospores of *S. subterranea* per ml buffer. Three replicate wells of microtitre plates were coated with extract of each microorganism for each MAb tested.

Limit of detection of zoospores The sensitivity of detection of zoospores was determined in indirect PTA-ELISA with alkaline phosphatase conjugate, using the MAb that recognized zoospores most strongly. A twofold dilution series was prepared of a suspension of live zoospores from 4 × 10^3 to 16 spores per ml carbonate buffer. 100 μl of each dilution, together with carbonate buffer only (zoospores absent), were placed in each of 12 replicate wells arranged as a randomized block design on two microtitre plates avoiding the outside wells, to which carbonate buffer only was added. A405 values were determined after 27 h of incubation at 37°C with enzyme substrate, and were subjected to an analysis of variance.

RESULTS AND INTERPRETATION

From the five fusions there were 3770 wells with hybridoma growth. Cell supernatant fluid from 259 (6.9%) of these wells contained antibodies that recognized zoospores but 254 (98.1%) also recognized tomato roots. Of the five hybridoma cell lines secreting MAbs that recognized zoospores but not the host, two also reacted weakly with plasmodia in comminuted roots and one of these also reacted weakly with cystosori. The antibody subclass of each MAb is presented in Fig. 4.1. The sensitivity of detection of plasmodia using MAbs 1 and 2 in ELISA was poorer than that obtained previously using a polyclonal antiserum from a rabbit (unpublished data). These MAbs would be of little use for quantification of plasmodia in roots of bait plants. Detection of cystosori using MAb 1 was also substantially poorer than that described by Harrison *et al.* (1993) using a polyclonal antiserum in ELISA.

There was little or no difference between the two enzyme systems for visualization of bound MAbs in sensitivity of detection of zoospores.

The 32 microorganisms other than *S. subterranea* against which the five MAbs were tested were either recognized only weakly (reaction < 10% of *S. subterranea* zoospores) or not at all.

The limit of detection of zoospores in buffer using MAbs 5 was *ca.* three spores per well (Fig. 4.2).

Statistical analysis showed that the mean A405 when there were 16 zoospores per ml was not significantly different from the buffer only control (zoospores absent). However, when there were 31 or more zoospores per ml mean A405 values were

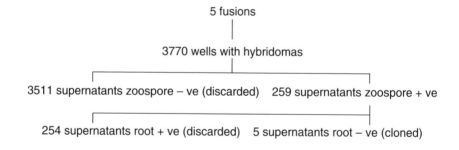

5 fusions

3770 wells with hybridomas

3511 supernatants zoospore – ve (discarded)   259 supernatants zoospore + ve

254 supernatants root + ve (discarded)   5 supernatants root – ve (cloned)

	MAb1	MAb2	MAb3	MAb4	MAb5
Origin: mouse no.	3	4	4	4	4
Antibody subclass	IgG1	IgG2b	IgG3	IgG2b	IgG3
Reaction with zoospores	++	++	++	++	+++
Reaction with plasmodia	+	+	-	-	-
Reaction with cystosori	+	-	-	-	-

Fig 4.1   Results from fusions and characteristics of MAbs.

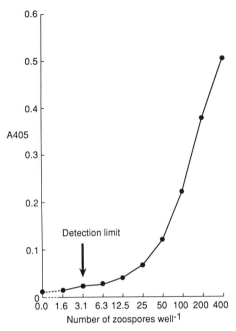

Fig 4.2. ELISA readings obtained with zoospores of *Spongospora subterranea* and MAb 5.

significantly greater than the control mean. Although A405 increased only slowly during incubation, by using many replicate wells of each dilution of zoospores in a randomized design on the microtitre plates and incubating with substrate over a prolonged period, we were able to detect low numbers of zoospores. Tests using all five MAbs alone or mixed in various combinations in PTA-ELISA, and further tests using MAb 5 only in DAS-ELISA did not improve on this sensitivity of detection.

The possibility of quantifying zoospores released from resting spores in field soil as a means of assessing contamination with *S. subterranea* is being investigated. This work is jointly supported by the Potato Marketing Board and the Scottish Office Agriculture and Fisheries Department.

REFERENCES

Elliott, C. G., Hendrie, M. R. and Knights, B. A (1966) The sterol requirements of *Phytophthora cactorum*. *Journal of General Microbiology* 42, 425–435.

Harrison, J. G., Rees, E. A., Barkers, H. and Lowe, R. (1993) Detection of spore balls of *Spongospora subterranea* on potato tubers by enzyme-linked immunosorbent assay. *Plant Pathology* 42, 181–186.

Jones, R. A. C. and Harrison, B. D. (1969) The behaviour of mop-top virus in soil, and evidence for its transmission by *Spongospora subterranea* (Wallr.) Lagerh. *Annals of Applied Biology* 63, 1–17.

Mohan, S. B. (1988) Evaluation of antisera raised against *Phytophthora fragariae* for detecting the red core disease of strawberries by enzyme-linked immunosorbent assay (ELISA). *Plant Pathology* 37, 206–216.

Wale, S. J. (1987) Powdery scab – are there any easy solutions? *Potato World* 4(4), 8–9.

5

Development of Monoclonal Antibody-based Immunological Assays for the Detection of Live Propagules of *Rhizoctonia solani* in the Soil

C. R. THORNTON,[1] F. M. DEWEY[1] AND C. A. GILLIGAN[2]

[1] *Department of Plant Sciences, University of Oxford, South Parks Road, Oxford OX1 3RB, UK;* [2] *Department of Plant Sciences, University of Cambridge, Downing Street, Cambridge CB2 3EA, UK.*

BACKGROUND

Rhizoctonia solani Kühn [Teleomorph: *Thanatephorus cucumeris* (Frank) Donk] is a ubiquitous soil-borne pathogen that attacks a wide range of crop plants including bean, beet, brassicas, potato and tomato as well as many herbaceous ornamentals. Common diseases include stem canker of potatoes and damping-off of many plant species. Various techniques, involving baiting and culture plating, have been developed for the qualitative and quantitative detection of *Rhizoctonia* species in the soil (Ko and Hora, 1971; Gangopadhyay and Grover, 1985; Trujillo *et al.*, 1987; Vincelli and Beaupré, 1989). Although quite effective, these conventional methods are labour intensive, cumbersome and require taxonomic expertise to differentiate species. Faster detection methods, such as immunoassays, that are specific and can easily be repeated, are needed to improve the precision by which populations of *R. solani* can be detected in soil.

Background to immunoassays for *R. solani*

Adams and Butler (1979) demonstrated that species-specific antibodies are present in polyclonal antisera raised against mycelial extracts of *R. solani*. The usefulness of these antibodies is masked, however, by the presence of many non-specific antibodies. Recently, Matthew and Brooker (1992) raised both polyclonal and monoclonal antibodies against proteins secreted *in vivo* from an anastomosis group 8 isolate of *R. solani*. Cross-reaction with three test organisms was minimal. However, they concluded that numerous other soil-borne fungi would need to be tested before discounting the

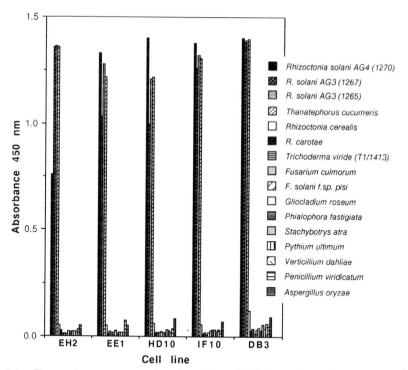

Fig 5.1.   Tissue culture supernatants from the cell lines *Rs.* EH2[3], EE1[2], HD10[2], IF10[1] and DB3[1] tested, by ELISA, against suface washings fromn *Rhizoctonia* spp. and other soil-borne fungi. Each bar represents the mean of duplicate samples.

possibility of cross-reaction with fungi other than *R. solani*. Dunsunceli and Fox (1992) using monoclonal antibodies developed for the detection of *Rhizoctonia* brown patch (Agri-Diagnostics Associates, Cinnamonson, NJ 08077, USA) claim to have developed an ELISA suitable for detecting the presence of a range of anastomosis groups of *T. cucumeris* in the soil. However, these antibodies are known to recognize both *R. solani* and *R. cerealis* (S. Miller, personal communication; Miller *et al.*, 1992).

This chapter describes the development of two immunological assays for the specific detection of live propagules of *R. solani* in the soil, a Diagnostic-ELISA and an immunofluorescence colony staining assay (IFC). These assays were developed for the qualitative and quantitative detection of *R. solani* in the soil. The Diagnostic-ELISA is currently being used to quantify the decay of *R. solani* inoculum in the soil based on quantal data from soil dilution series.

Production of monoclonal antibodies

Monoclonal antibodies (MAbs) were raised against an anastomosis group 4 isolate of *R. solani*. Mice were immunized using either phosphate buffered saline (PBS) suspensions of lyophilized mycelium plus Quil A adjuvant or with a solubilized acetone precipitate prepared from cell-free surface washings from 5-day-old PDA solid slant cultures grown at 17°C. Hybridoma supernatants were screened, by ELISA, against surface washings from a range of soil-borne fungi (Fig. 5.1). Four of the cell lines raised, *Rs.*EH2[3], EE1[2],

HD10[2] and IF10[1] produce MAbs that are species-specific. They recognize, by ELISA and immunofluorescence (IF), antigens from *R. solani* but do not recognize other related or unrelated species of soil-borne fungi. The remaining cell line *Rs*.DB3[1] produces MAbs that cross-react slightly, by ELISA, with antigens from *R. cerealis* but do not recognize other related or unrelated soil-borne fungi, by ELISA. Indirect evidence from heat, protease and periodate treatment of the antigens indicates that MAbs from the cell lines *Rs*.EE1[2], HD10[2], IF10[1] and DB3[1] recognize carbohydrate epitopes while the remaining cell line *Rs*.EH2[3] recognizes protein epitopes, on a glycoprotein. The four MAbs that recognize carbohydrate epitopes belong to the subclass IgM while the MAb that recognizes protein epitopes is of the IgA subclass. MAbs from the cell line *Rs*.EE1[2] have been used to develop Diagnostic-ELISA and immunofluorescence colony staining (IFC) immunoassays for the detection of *R. solani* in the soil.

PROTOCOL DIAGNOSTIC-ELISA

Antigen extraction

Material

1. Sterile scintillation polyvials (Type-900, C and G Chemicals and Services Ltd).
2. Sterile aqueous semi-selective medium (SSM): 1 l of tap water, 0.4 g gallic acid, 50 mg streptomycin sulphate, 50 mg chloramphenicol, 2.0 g $NaNO_3$, 1.0 g KH_2PO_4, 0.5 g $MgSO_4.7H_2O$, 0.01 g $FeSO_4.7H_2O$, 0.5% (w/v) potato starch (BDH) and 1.38 ml of Filex (Fisons plc, Ipswich) diluted to 10 μg ml^{-1} of active ingredient. The medium is sterilized, prior to the addition of streptomycin sulphate and Filex, by autoclaving at 121°C for 15 min.
3. 1.5 ml polypropylene micro-centrifuge tubes (Scotlab Ltd).

Method

1. Place 1 g samples of test soil in individual sterile polyvials and add 1 ml of sterile aqueous SSM.
2. Incubate in the dark on a Spiramix rotary mixer, at 25°C, for 48 h.
3. Transfer the sludge to 1.5 ml micro-centrifuge tubes.
4. Centrifuge for 2 min at 13,000g.
5. Transfer the supernatants to clean micro-centrifuge tubes.
6. Heat the extracts to 100°C for 5 min.
7. Allow the extracts to cool and then store at 4°C until assayed.

ELISA protocol

Materials

1. 96-well Immulon II flat-bottomed microtitre plates (Dynatech Labs., Inc.).
2. Washing buffer and diluent for secondary horseradish peroxidase conjugate: Tris–casein buffer (TCB); 154 mM NaCl, 0.3% (w/v) casein (BDH), 10 mM Tris/HCl, pH 7.6).
3. Goat anti-mouse polyvalent immunoglobulins (IgG, IgM, IgA) peroxidase conjugate (Sigma A-0412).

4. Substrate buffer; 5 ml 0.2 M sodium acetate, 195 μl 0.2M citric acid, 5 μl H_2O_2, 5 ml H_2O and 100 μl 3, 3, 5, 5-tetramethylbenzidine (TMB) stock solution (10 mg TMB in 1 ml dimethyl sulphoxide (DMSO), stored at 4°C).

Method

1. Coat microtitre wells with soil extracts (see Antigen extraction above), 100 μl per well.
2. After incubating the wells overnight at 4°C, invert and flick out residual extract.
3. Wash three times with TCB, 1 min each time.
4. Block each well with 200 μl TCB for 10 min.
5. Incubate wells with 100 μl serum-free monoclonal antibody supernatant from the cell line *Rs*.EE1[2], containing 0.3% (w/v) casein, for 1 h.
6. Wash four times with TCB, 1 min each time.
7. Incubate the wells with 100 μl secondary antibody peroxidase conjugate diluted 1/500 in TCB for 1 h.
8. Wash four times with TCB, 1 min each time.
9. Rinse briefly with distilled water.
10. Add 100 μl of substrate solution to each well and incubate for 15 min.
11. Stop the reaction by the addition of 100 μl 3 M H_2SO_4.
12. Read absorbance values at 450 nm on an ELISA-reader.

NB All incubation steps are performed at 37°C, with shaking.

RESULTS

Sensitivity and high background levels presented serious problems in developing an ELISA test for the detection of the pathogen in both naturally and artificially infested soil samples. However, by incorporating a biological amplification step followed by centrifugation and heating of soil extracts, we have increased the level of antigen available for detection, have ensured the detection of only live propagules and have reduced background interference to a negligible level. Using this assay system, with supernatants from the cell line EE1[2], we obtained high absorbance values for artificially infested soil (CPS+) extracts (Fig. 5.2). Similar absorbance values were obtained with the positive standards.

PROTOCOL IMMUNOFLUORESCENCE COLONY-STAINING (IFC)

Materials

1. Washing and incubation buffer: phosphate buffered saline (PBS); 0.8% NaCl, 0.02% KCl, 0.115% Na_2HPO_4, 0.025% KH_2PO_4, pH 7.2.
2. Goat anti-mouse IgG (whole molecule) FITC conjugate (Sigma F-9006).
3. Sterile solid semi-selective medium; 1 l of tap water, 0.4 g gallic acid, 50 mg streptomycin sulphate, 50 mg chloramphenicol, 2.0 g $NaNO_3$, 1.0 g KH_2PO_4, 0.5 g $MgSO_4.7H_2O$, 0.01 g $FeSO_4.7H_2O$, 0.5% (w/v) potato starch (BDH) and 1.38 ml of Filex (Fisons plc, Ipswich) diluted to 10 μg ml^{-1} of active ingredient and 1.2% (w/v) agar no.3 (Oxoid). The medium is sterilized, prior to the addition of streptomycin sulphate and Filex, by autoclaving at 121°C for 15 min.
4. Citifluor mounting medium (Citifluor Ltd, City University, London).

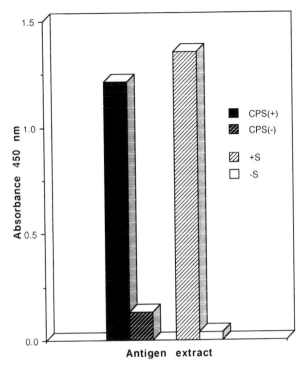

Fig 5.2. Supernatant from the cell line *Rs*. EE1² tested, by Diagnostic–ELISA, against extracts from matrices artificially infested with *R. solani*.

Key. CPS(+) extract from chopped potato soil (artificially infested with *R. solani* AG4 isolate 1270); CPS(−) extract from chopped potato soil (uninfested); +S positive standard; filtrate from shake cultures of *R. solani* (1270) grown in aqueous semi-selective medium; −S negative standard; uninoculated semi-selective medium.

Each bar represents the means of four replicate values corrected by subtracting values obtained using wells probed with serum-free tissue culture media alone.

Method

1. Incubate soil samples on solid SSM contained in sterile Petri dishes, for 24 h at 25°C.
2. Dry the agar, *in situ*, into thin films by warming in an oven at 40°C for 24 h.
3. Excise the colonies and immobilize on to glass slides using double-sided Sellotape.
4. Incubate the colonies with *Rs*.EE1² MAb supernatant, containing fetal calf serum, for 1 h at room temperature.
5. Wash the slides with excess PBS (3 × 5 min rinses).
6. Incubate the colonies with FITC conjugate diluted 1/40 in PBS for 30 min at room temperature.
7. Wash with excess PBS (3 × 5 min rinses).
8. Mount the samples in Citifluor and store in the dark at 4°C.
9. Examine the colonies using epifluorescence with a UV excitation filter of 365 nm and an absorption filter of 420 nm.

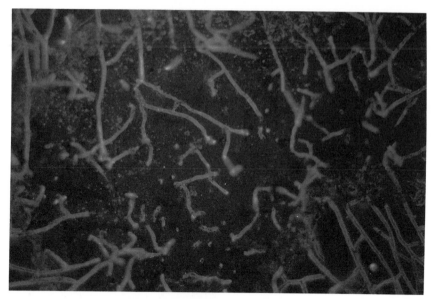

Fig 5.3. Photomicrograph of *R. solani* isolated, on solid semi-selective medium, from soil artificially infested with *R. solani* as chopped potato inoculum and stained using the immunofluorescence colony-staining (IFC) technique. (× 410 magnification).

RESULTS

The medium used to isolate *R. solani* was found to be only partially selective and has therefore been referred to throughout as a semi-selective medium. It supported the growth of a number of other soil fungi notably *R. cerealis, Trichoderma viride* and species of *Penicillium*. The growth of these other soil fungi on the plates made discrimination between colonies of *R. solani* and *R. cerealis* very difficult under white light using bright field optics even after incubation for only 24 h. However, when such plates were immunostained with supernatant from the cell line *Rs*.EE1[2], colonies of *R. solani* fluoresced brightly under UV light while those of *R. cerealis* did not. Immunofluorescent positive colonies were found in cultures from artificially infested matrices. No positive colonies were recorded from uninfested control soil (CPS−). An example of a positive fluorescent colony is given in Fig. 5.3.

LIMITATIONS

The cost of commercial fluorescent conjugates prohibits extensive use of the IFC immunoassay for *R. solani* detection in the soil. However, this method is extremely useful in cases where differentiation between *R. solani* and other *Rhizoctonia* species is not possible by conventional means.

AVAILABILITY OF *RHIZOCTONIA SOLANI* SPECIFIC MONOCLONAL ANTIBODIES

Contact Dr F. M. Dewey, Department of Plant Sciences, South Parks Road, Oxford, OX1 3RB, UK.

REFERENCES

Adams, G. C. and Butler, E. E. (1979) Serological relationships among anastomosis groups of *Rhizoctonia solani*. *Phytopathology* 69(6), 629–633.

Dusunceli, F. and Fox, R. T. V. (1992) Accuracy of methods for estimating the size of *Thanatephorus cucumeris* populations in soil. *Soil Use and Management* 8(1), 21–26.

Gangopadhyay, S. and Grover, R. K. (1985) A selective medium for isolating *Rhizoctonia solani* from soil. *Annals of Applied Biology* 106, 405–412.

Ko, W-Hsuing and Hora, F. K. (1971) A selective medium for the quantitative determination of *Rhizoctonia solani* in soil. *Phytopathology* 61, 707–710.

Matthew, J. S. and Brooker, J. D. (1991) The isolation and characterisation of polyclonal and monoclonal antibodies to anastomosis group 8 of *Rhizoctonia solani*. *Plant Pathology* 40, 67–77.

Miller, S. A., Rittenburg, J. H., Petersen, F. P. and Grothaus, G. D. (1992) From the research bench to the market place: development of commercial diagnostic kits. In: Duncan, J. M. and Torrance, R. (eds) *Techniques for the Rapid Detection of Plant Pathogens*. Blackwell Scientific, Oxford, pp. 208–221.

Trujillo, E. E., Cavin, C. A., Aragaki, M. and Yoshimura, M. A. (1987) Ethanol–potassium nitrate medium for enumerating *Rhizoctonia solani*-like fungi from soil. *Plant Disease* 71(12), 1098–1100.

Vincelli, P. C. and Beaupré, C., M. S. (1989) Comparison of media for isolating *Rhizoctonia solani* from soil. *Plant Disease* 73, 1014–1017.

6

Differentiation Between *Fusarium culmorum* and *Fusarium graminearum* by RFLP and with Species-specific DNA Probes

B. Koopmann, P. Karlovsky and G. Wolf

Institute of Plant Pathology and Plant Protection, University of Göttingen, Grisebachstrasse 6, D-37077 Göttingen, Germany.

Background

The genus *Fusarium* represents a heterogeneous group of fungi encompassing many important plant pathogens. The taxonomy of *Fusarium* is based on morphological characters including the presence or absence, the shape and the dimensions of microconidia and macroconidia, macroconidial basal cells and chlamydospores; in addition, growth and colour development on different media are used as markers in practice. These characters are not stable and can be altered readily under different growing conditions and by the occurrence of culture mutations (Bilai, 1970). The difficulty with classical taxonomic markers in *Fusarium* is well illustrated by the fact that the system introduced by Wollenweber and Reinking (1935) and further developed by Gerlach (Gerlach, 1970; Gerlach and Nirenberg, 1982) distinguishes 16 sections encompassing about 70 species, while the rival system of Snyder and Hansen (1945), continued by Messiaen and Cassini (1968), emphasizes the within-species variability and recognizes only nine species of *Fusarium* in all.

Fusarium culmorum and *F. graminearum* are serious cereal pathogens. They infect seedlings, roots, stem base and ears. The economic importance of these two species is further enhanced by their ability to synthesize a number of mycotoxins harmful to men and animals (e.g. zearalenone, fusarins, trichothecenes). The current method of differentiation between *F. culmorum* and *F. graminearum* relies on highly variable morphological traits. Since a reliable diagnosis is a prerequisite for efficient plant protection, there is demand for a simple and reliable method for the detection and differentiation of *Fusarium* species that does not use morphological characters. Both immunology and DNA technology offer methods to this end. The advantage of DNA methods and especially of the polymerase chain reaction, is high sensitivity and specificity.

DNA hybridization probes of two kinds may be used to differentiate between two species: (i) RFLP probes that possess sequence similarity to genomic segments in both species but reveal different banding patterns in Southern hybridization; and (ii) genuine species-specific DNA probes representing sequences that are present in the genome of one species but absent in the other one. The advantages of the latter probes are the possibility to use them as diagnostic techniques without electrophoretic separation (i.e. by dot hybridization) and the fact that they are supposed to be very reliable. While the value of an RFLP probe as a diagnostic tool is jeopardized by any single point mutation within one of the boundary restriction sites and by the occurrence of new restriction sites within the restriction fragment detected on the gel, the hybridization result obtained with a species-specific DNA sequence is not influenced by point mutations.

Isolation of a species-specific DNA hybridization probe for *Fusarium culmorum*

The objective of this work was to clone a fragment from the genome of *Fusarium culmorum* that would not hybridize with genomes of fungi known to occur in the natural habitat of *F. culmorum*, in particular *F. graminearum*. Two strategies were used to this end: (i) shotgun cloning of *Sau*3A fragments of *F. culmorum* DNA in pUC19 (Yanisch-Perron *et al.*, 1985) cleaved with *Bam*HI and screening of individual clones by hybridization against nuclear genomes of *F. culmorum* and *F. graminearum*; and (ii) cloning of fragments enriched for specific *F. culmorum* sequences by hybridization with an excess of *F. graminearum* genomic DNA in solution followed by the separation of hybrids from unhybridized DNA. This contribution describes the results obtained with 20 randomly selected single-copy Sau3A fragments.

Twenty probes from the genome of *F. culmorum* SAGW29 were classified according to the hybridization pattern they produced with genomic DNA of *F. culmorum* SAGW29 and *F. graminearum* SAGW3 digested with *Eco*RI (Table 6.1). The high proportion of probes hybridizing with genomic digests of both species, half of them even producing identical hybridization patterns, indicates that both genomes are very similar on the nucleotide sequence level. Probes detecting RFLP between the species will be described in detail later.

The only probe found to be specific for the genome of *F. culmorum* SAGW29 (plasmid 220–24) was used for hybridization with DNA from *F. culmorum* and *F. graminearum* strains and from other fungi (see Results) in order to test its suitability as a diagnostic probe for *F. culmorum* in the presence of other fungal pathogens of cereals.

Protocols

Hybridization assay

Materials

1. Fungal isolates stored in sterile soil at 4°C.
2. Czapek-Dox medium (E. Merck, 1983), SNA medium (Nirenberg, 1976).
3. Standard DNA buffers: TE, standard saline phosphate EDTA (SSPE) and H-buffer for restriction enzymes according to Sambrook *et al.* (1989).
4. Lyophilisator, mortars, sand and buffers for DNA isolation according to Möller *et al.* (1992).
5. Spermine hydrochloride 0.1 M, exchange buffer (0.3 M K-acetate, 10 mM Mg-acetate) and ethanol for DNA purification.

6. Agarose, TAE buffer (Sambrook *et al.*, 1989), apparatus for horizontal electrophoresis, ethidium bromide $0.5\,\mu g\,ml^{-1}$, UV transilluminator, Polaroid camera.
7. HCl $0.25\,M$, NaOH $0.4\,M$, nylon membranes Hybond N^+ (Amersham), VacuGene blotting apparatus (Pharmacia/LKB), vacuum oven 110°C.
8. Polymerase buffer: 0.2 mM of each dATP, dCTP and dGTP, 0.17 mM dTTP, 0.03 mM digoxigenen-11-dUTP (Boehringer), 2 mM $MgSO_4$ (for Vent polymerase) or 1.5 mM $MgCl_2$ (for *Taq* polymerase); mineral oil, chloroform; universal M13/pUC primers GTAAAACGACGGCCAGT and CAGGAAACAGCTATGAC; Vent polymerase (New England Biolabs) and *Taq* polymerase (Appligene); OmniGene thermocycler with tube sensor (Hybaid).
9. Solutions for DNA hybridization and detection according to the DIG DNA Labelling and Detection Kit (Boehringer), hybridization oven (Hybaid), AMPPD and Lumigen PPD (Boehringer), Fuji RX film.

Methods

DNA isolation Fungi were grown in Czapek-Dox or in SNA medium at RT with agitation (200 ml in 1000 ml Erlenmeyer flasks). Total DNA from lyophilized mycelium was prepared according to the method of Möller *et al.* (1992). Some preparations obtained in this way were cleaved inefficiently or not at all by restriction enzymes. After purification by one spermine precipitation, all DNA samples could be digested under standard conditions, so we included spermine precipitation as a final step in all our DNA preparations. The procedure is a modification of the method described by Hoopes and McClure (1981): DNA in TE was mixed with 0.15 vol. 100 mM spermine hydrochloride. The pellet formed within 1 to 10 min at RT was centrifuged for 5 min (according to the tubes used at 5000 to 14,000 g) and dissolved in 1 vol. exchange buffer with agitation (2–4 h at 50°C or overnight at RT). DNA was precipitated from the exchange buffer with 3 vol. ethanol, washed with 70% ethanol, dried and dissolved in TE.

Electrophoresis and Southern blotting Total fungal DNA was completely digested with *Eco*RI and electrophoretically separated on a 0.8% horizontal agarose gel in Tris-acetate buffer using standard conditions (Sambrook *et al.*, 1989). After staining with ethidium bromide and photographing on a UV transilluminator, DNA was depurinated by agitating the gel in 0.25 M HCl for 5 min. The gel was equilibrated in 0.4 M NaOH for 15 min and the transfer to positively charged nylon membranes Hybond N^+ (Amersham) was per-

Table 6.1. Southern hybridization of randomly selected single copy Sau3A fragments from *F. culmorum* SAGW29 with the *Eco*RI digest of *F. culmorum* SAGW29 and *F. graminearum* SAGW3 genomic DNA.

Group	Hybridization specifity	Number of probes
I	Identical pattern with both species (= unspecific probes)	10
II	Different patterns with *F. culmorum* and *F. graminearum* (differentiation possible only on the basis of RFLP)	9
III	Positive signal with *F. culmorum*, no signal with *F. graminearum* (species specific probe)	1

1 2 3 4 5 6 7 8 9 10 11 12 13 14 15 16 17 18 19

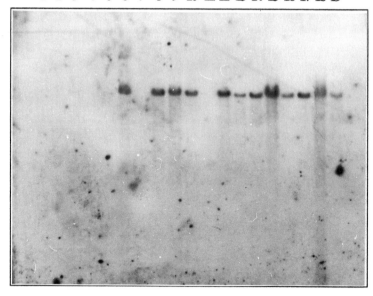

Fig 6.1. Southern hybridization of *Eco*RI DNA digests of *Fusarium* isolates with probe 220–24. Lines 1 to 7, *F. graminearum* isolates SAGW191, SAGW103, SAGW102, SAGW101, SAGW100, SAGW6 and SAGW3; lines 8 to 19, *F. culmorum* isolates SAGW29, SAGW7, SAGW25, SAGW82, SAGW114, SAGW115, SAGW116, SAGW188, CBS93, DSM62184, DSM62223 and SAGW2.

formed using vacuum blotting technique (VacuGene, Pharmacia) at 40 mbar reduced pressure for 90 min with 0.4 M NaOH as transfer medium. The membranes were washed in 2 × SSPE and baked for 1 h at 110°C.

Labelling of probes with digoxigenin by PCR PCR was used to label inserts selectively by amplifying them from universal primers embracing the multiple cloning site of the vector (see Materials). Plasmid DNA (2–4 ng) linearized with *Pvu*II was used as template. The reaction mixture containing 15 pmol of each primer in 15 μl polymerase buffer was covered with 20 μl mineral oil and loaded into a thermocycler with a tube sensor (OmniGene, Hybaid). After initial denaturation at 95°C for 150 s, the mixture was cooled to 85°C, polymerase was added under the oil layer with gentle mixing (0.15 u Vent or 0.45 u *Taq*) and the first annealing step at 55°C for 40 s was performed. Then, 20 cycles were run, consisting of 15 s polymerization at 72°C, 40 s denaturation at 95°C and 40 s annealing at 55°C each. The terminal extension was performed for 5 min at 72°C. The separation of the reaction mixture from mineral oil was facilitated by adding 50 μl chloroform. An aliquot of the product was checked on an electrophoretic gel before the amplified DNA was used as a hybridization probe. The mobility of the digoxigenin-labelled DNA fragments in agarose gels was reduced as compared to amplification products obtained in a control reaction with digoxigenin-11-dUTP replaced by an equimolar amount of dTTP. Increasing the digoxigenin-11-dUTP concentration to 0.06 mM in line with Boehringer protocol did not improve the sensitivity of detection.

Hybridization and chemiluminescent detection PCR products were used as hybridization probes without purification, the hybridization was performed in 50% formamide at 42°C

(Sambrook *et al.*, 1989). After washing two times for 15 min in 2 × SSPE, 0.1% SDS at RT and two times for 15 min at 55°C in 0.2× SSPE, 0.1% SDS, the bands hybridizing with the probe were detected with the help of anti-digoxigenin Fab-fragments conjugated to alkaline phosphatase essentially as recommended in the documentation to the DIG DNA Labelling and Detection Kit and to the AMPPD-substrate (Boehringer), except that the substrate (AMPPD or Lumigen PPD) was used at a concentration of $10 \mu g\,ml^{-1}$ instead of $100 \mu g\,ml^{-1}$. Filters were exposed to Fuji RX films for 0.5 to 4 h.

RESULTS

The fragment cloned in plasmid 220–24 was labelled with digoxigenin and used as a hybridization probe for Southern blots with 31 isolates of *Fusarium culmorum*, 11 isolates of *F. graminearum* and 20 isolates of seven species of plant pathogenic fungi associated with roots and stems of cereals (see Table 6.2 for sources of isolates). The results of these experiments are shown at Figs 6.1 to 6.4.

Of 31 *F. culmorum* strains tested, 28 revealed the same band (5.30 kb) as the strain SAGW29 from which the probe was isolated. Among three strains not hybridizing with the probe, the strain SAGW82 was a single-spore isolate from SAGW2. Since the parental strain hybridized as expected (Fig. 6.1, line 19), the loss of the band in SAGW82

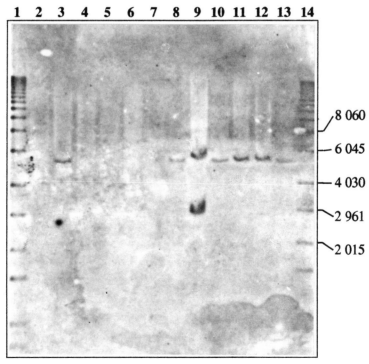

Fig 6.2. Southern hybridization of *Eco*RI DNA digests of *Fusarium* isolates with probe 220–24. Lines 1 and 14, size marker (length in bp indicated on the right margin); lines 2 to 7, *F. graminearum* isolates SAGW100, SAGW102, 67638, 67639, 67640 and 64221; lines 8 to 13, *F. culmorum* isolates H2, H3, H19, H27, H29 and H30.

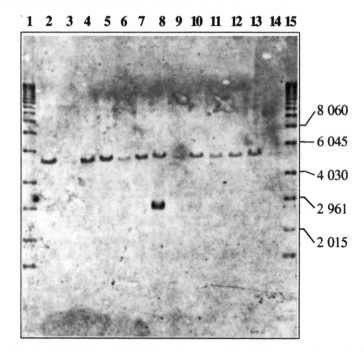

Fig 6.3. Southern hybridization of *Eco*RI cleaved genomic DNA of *Fusarium culmorum* isolates with probe 220–24. Lines 1 and 15, size marker (length in bp indicated on the right margin); lines 2 to 14, isolates H33, H41, H43, H49, H55, H62, H66, H67, H69, H70, H71, H72 and H73.

may be explained either as a result of a genome rearrangement or as a contamination with another organism arising in the process of growing single-spore isolates from SAGW2 in the laboratory and mistaken for *F. culmorum*. Since another DNA probe, differentiating between *F. culmorum* and *F. graminearum* by RFLP, did not produce any band with SAGW82, and on the basis of the pattern of repetitive DNA fragments (unpublished results), we incline to the second hypothesis. Microscopical re-examination confirmed this conclusion (M. Wegener, Göttingen, 1993, personal communication). Another two strains declared as *F. culmorum* and not hybridizing with 220–24 were H41 and H73 (Fig. 6.3, lines 3 and 14). These negative results with H41 may be explained by the low amount of this DNA on the gel.

Strain DSM62223 is described as *F. graminearum* in the DSM catalogue. The strain was reclassified as *F. culmorum* on the basis of both molecular and morphological characters in two different laboratories (E. Möller, personal communication) before we did these experiments.

In addition to the diagnostic band of 5.3 kb, a new band migrating at 2.88 kb was found in DNA digests of two isolates of *F. culmorum*: H3 from Norway (Fig. 6.2, line 9) and H66 from Italy (Fig. 6.3, line 8). The intensity of the band indicated that it contained more copies of a hybridizing fragment than the 5.3 kb band. Since the mobilities and the relative intensities of hybridizing bands in both strains were identical, they seem to be the result of the same amplification event. Further investigation (DNA fingerprinting) should reveal whether this amplification happened twice independently or whether the two strains inherited the amplified DNA from a common ancestor. The geographic origin of the strains and the fact that amplification of a sequence homologous to the probe was not found in any other *F. culmorum* strain investigated are arguments against the common ancestor hypothesis.

Among 11 *F. graminearum* strains, two strains revealed a band typical of *F. culmorum*. Microscopical re-examination revealed that the SAGW6 strain (Fig. 6.1, line 6) possessed morphological characters typical of *F. culmorum* (M. Wegener, personal communication). So far we have no morphological results on the second strain with an anomalous hybridization, SAGW102 (Fig. 6.1, line 3; Fig. 6.2, line 3).

The only DNA hybridizing with 220–24 apart from *F. culmorum* and *F. graminearum* preparations was the DNA of *Fusarium avenaceum* DSM62161 (Fig. 6.4, line 9). This hybridization signal was proved to be an artefact, probably due to the contamination of the sample by the 220–24 plasmid: Neither other *F. avenaceum* isolates (Fig. 6.4, lines 10 und 11) nor the repetition of the experiment with DSM62161 (data not shown) produced a hybridization signal with 220–24.

We conclude that probe 220–24 is suitable for specific detection of *F. culmorum*. The absence of sequence homologies to the probe in genomes of *F. graminearum* and of seven other fungal species sharing their natural habitat with *F. culmorum* guarantees the reliability of the hybridization assay. In contrast to the application of species-specific RFLP probes, the diagnostic value of the assay based on probe 220–24 is independent of point mutations creating or destroying restriction sites.

1 2 3 4 5 6 7 8 9 10 11 12 13 14 15 16 17 18 19 20 21 22

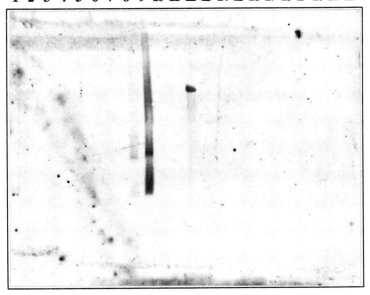

Fig 6.4. Southern hybridization of *Eco*RI genomic digests. Lines 1 and 2, *Fusarium sulphureum* isolates SAGW71 and SAGW72; lines 3 to 5, *Fusarium oxysporum* isolates SAGW124, SAGW125 and SAGW138; lines 6 to 8, *Microdochium nivale* isolates SAGW78, SAGW186 and SAGW200; lines 9 to 11, *Fusarium avenaceum* isolates DSM62161, SAGW39 and CBS95; line 12, *Fusarium culmorum* SAGW29; line 13, *Fusarium graminearum* SAGW3; lines 14 to 16, *Pseudocercosporella herpotrichoides* isolates SAGW10R, SAGW14R and INRA164; lines 17 to 19, *Rhizoctonia cerealis* isolates SAGW106, SAGW105 and INRA161; lines 20 to 22, *Septoria nodorum* isolates SAGW31, INRA156 and SAGW214.

Table 6.2. Origin and source of fungal isolates.

Species	Isolate number	Host/habitat	Geographic location	Country	Year	Source
Fusarium	CBS 95	—	—	—	—	1
avenaceum	DSM 62161	—	—	—	—	2
	SAGV 39	Triticum aestivum	Göttingen	D	1985	7
Fusarium	CBS 93	—	—	—	—	1
culmorum	DSM 62223	—	—	—	—	2
	DSM 62184	—	—	—	—	2
	H 2	Hordeum vulgare		SF	1982	3
	H 3	Triticum aestivum		N	1982	3
	H 19	Triticum durum	Hohenheim	D	1987	3
	H 27	Secale cereale	Bergen	D	1991	3
	H 29	Triticum durum	Hohenheim	D	1991	3
	H 30	Triticum dicoccum	Vaihingen	D	1991	3
	H 33	Secale cereale	Chewendowa	PL	1991	3
	H 41	—	Szeged	H	1991	3
	H 42	Triticum aestivum	Szeged	H	1991	3
	H 43	Triticum aestivum	Szeged	H	1991	3
	H 49	—	Wageningen	NL	1987	3
	H 55	Triticum spelta	Rottenburg	D	1988	3
	H 62	Secale cereale	Svalöf	S	1992	3
	H 66	Triticum aestivum	—	I	1992	3
	H 67	Triticum aestivum	—	I	1992	3
	H 69	—	Limburg	NL	1988	3
	H 70	—	Ellinghausen	CH	1989	3
	H 71	Zea mays	New South Wales	AUS	1992	3
	H 72	Soil	Queensland	AUS	1992	3
	H 73	Soil	Tasmania	AUS	1992	3
	SAGW 2	—	—	—	—	6
	SAGW 7	Triticum aestivum	Wülfingen	D	1985	7
	SAGW 25	Triticum aestivum	Rosdorf	D	1985	7
	SAGW 29	Triticum aestivum	Rosdorf	D	1985	7
	SAGW 82	—	—	—	—	7
	SAGW 114	Pisum sativum	Oldenburg	D	1987	7
	SAGW 115	Pisum sativum	Oldenburg	D	1987	7
	SAGW 116	Pisum sativum	Oldenburg	D	1987	7
	SAGW 118	Triticum aestivum	Söhlingen	D	1987	10
Fusarium	64221	Hordeum vulgare	—	SF	—	5
graminearum	67638	Triticum aestivum	Bavaria	D	1992	5
	67639	Triticum aestivum	—	F	1992	5
	67640	Triticum aestivum	—	F	1992	5
	SAGW 3	—	—	—	—	6
	SAGW 6	—	—	—	—	7
	SAGW 100	Triticum aestivum	Harrissa	GR	1987	7
	SAGW 101	Triticum aestivum	Harrissa	GR	1987	7
	SAGW 102	Triticum aestivum	Harrissa	GR	1987	7
	SAGW 103	Triticum aestivum	Harrissa	GR	1987	7
	SAGW 191	Secale cereale	Hohenheim	D	1987	10
Fusarium	SAGW 124	—	Aalsmeer	NL	1987	7
oxysporum	SAGW 125	—	Baarn	NL	1987	7
	SAGW 138	—	Morocco	MA	1987	9

Table 6.2. *(continued)*

Species	Isolate number	Host/habitat	Geographic location	Country	Year	Source
Fusarium	SAGW 71	—	—	—	—	7
sulphureum	SAGW 72	—	—.	—	—	7
Microdochium	SAGW 78	Secale cereale	Kassel	D	—	7
nivale	SAGW 186	Secale cereale	Schwäbische Alb	D	1987	10
	SAGW 200	Apera spica venti	Göttingen	D	1990	7
Pseudocerco-	INRA 164	—	Ille et Vilaine	F	1986	4
sporella	SAGW 10R	—	—	—	1991	8
herpotrichoides	SAGW 14R	—	—	—	1991	8
Rhizoctonia	INRA 161	—	Marne	F	1981	4
cerealis	SAGW 105	Triticum aestivum	Loxstedt-Dünenfähr	D	1984	7
	SAGW 106	—	Schaardeich	D	1984	7
Septoria	INRA 156	—	Aisne	F	1988	4
nodorum	SAGW 31	—	Horste	D	1976	7
	SAGW 214	Triticum aestivum	—	D	—	7

Source:
1 Centraal Bureau voor Schimmelculturen Schimmelculteren, NL.
2 Deutsche Sammlung von Mikroorganismen, D.
3 Landessaatzuchtanstalt, Universität Hohenheim, Dr Miedaner, D.
4 Institut National de la Recherche Agronomique, F.
5 Biologische Bundesanstalt Berlin, Dr Nirenberg, D.
6 Pflanzenschutzamt Mainz, D.
7 Institut für Pflanzenpathologie, Universität Göttingen, D.
8 Schering AG, Berlin, D.
9 Institut Régional de la Recherche Agronomique, SCAS Marrakech, MA.
10 Landessaatzuchtanstalt, Universität Hohenheim, Dr Moser, D.

LIMITATIONS

The assay presented is based on Southern hybridization. This procedure requires microgram amounts of DNA purified to a level allowing restriction digestion. The assay includes three time-consuming operations: gel electrophoresis, blotting to nylon membranes and hybridization. The application of a dot blot hybridization procedure would result in a significant simplification of the experimental protocol, but the requirement of a rather high amount of pure DNA would still constitute a severe limitation for routine use. We plan to determine the nucleotide sequence of the *F. culmorum*-specific probe 220–24 and to design specific primers for PCR on its basis. The application of PCR should enable the detection and possibly quantification of small amounts of *F. culmorum* biomass in field samples containing an excess of DNA originating from soil microorganisms and/or host plants.

Acknowledgements

We are obliged to Drs Th. Miedaner (University Stuttgart) for providing us with *Fusarium* strains and E. Möller (University of Stuttgart) and M. Wegener (University of Göttingen) for communicating unpublished results.

References

Bilai, V. I. (1970) Experimental morphogenesis in the fungi of the genus *Fusarium* and their taxonomy. *Ann. Acad. Sci. Fenn. A, IV Biologica* 168, 7–18.

Gerlach, W. (1970) Suggestions to an acceptable modern *Fusarium* system. *Ann. Acad. Sci. Fenn. A, IV Biologica* 168, 37–49.

Gerlach, W. and Nirenberg, H. (1982) The genus *Fusarium* – a pictorial atlas. *Mitteilungen der Biologische Bundesanstalts für Land- und Forstwirtschaft* 209, 1–406.

Hoopes, B. C. and McClure, W. R. (1981) Studies on the selectivity of DNA precipitation by spermine. *Nucleic Acids Research* 9, 5493–5504.

Merck, E. (1983) *Handbuch Nährböden MERCK.* E. Merck, Darmstadt.

Messiaen, C. M. and Cassini, R. (1968) Recherches sur les fusarioses. IV. La systématique des *Fusarium. Annales des Epiphyties.* 19, 387–454.

Möller, E. M., Bahnweg, G., Sandermann, H. and Geiger, H. H. (1992) A simple and efficient protocol for isolation of high molecular weight DNA from filamentous fungi, fruit bodies, and infected plant tissues. *Nucleic Acids Research* 20, 6115–6116.

Nirenberg, H. (1976) Untersuchungen über die morphologische und physiologische Differenzierung in der Fusarium-Sektion Liseola. *Mitteilungen der Biologische Bundesanstalts für Land- und Forstwirtschaft Berlin-Dahlem* 169, 1–117.

Sambrook, J., Fritsch, E. F. and Manitatis, T. (1989) *Molecular Cloning: A Laboratory Manual.* Cold Spring Harbor Laboratory Press, Cold Spring Harbor.

Snyder, W. C. and Hansen, H. N. (1945) The species concept in *Fusarium* with reference to Discolor and other sections. *American Journal of Botany* 32, 657–666.

Wollenweber, H. W. and Reinking, O. A. (1935) *Die Fusarien, ihre Beschreibung, Schadwirkung und Bekämpfung.* Paul Parey, Berlin.

Yanisch-Perron, C., Vieira, J. and Messing, J. (1985) Improved M13 phage cloning vectors and host strains: nucleotide sequence of the M13mp18 and pUC19 vectors. *Gene* 33, 103–119.

7

RAPDs of *Fusarium culmorum* and *F. graminearum*: Application for Genotyping and Species Identification

A. G. Schilling, E. M. Möller and H. H. Geiger

Institute of Plant Breeding, Seed Science and Population Genetics, University of Hohenheim, PO Box 700562/350, D-70593 Stuttgart, Germany.

Background

The closely related species *Fusarium culmorum* (W. G. Smith) Sacc. (teleomorph unknown), and *F. graminearum* Schwabe [teleomorph *Gibberella zeae* (Schw.) Petch] are worldwide fungal pathogens causing foot rot and head blight of cereals and grasses (Nelson *et al.*, 1981). Plants are infected by seed- or soil-borne inocula produced on debris of cereals. Early infections cause seedling death, or later, stem lesions are formed which can lead to foot rot disease. Affected stems lose stability and functioning of the vascular tissue which result in preharvest lodging and seed ripening (Nelson *et al.*, 1981). Visual diagnosis is often aggravated or even impossible due to the presence of other foot rot fungi such as *Pseudocercosporella herpotrichoides*, *Microdochium nivale* and *F. avenaceum*. Head infections are caused by water splashing of conidiospores from the crown up to the stem and leaf sheaths or directly by wind-dispersal of spores. Infection of the heads during flowering will lead to bleached spikelets. As a consequence, seed set fails or only shrivelled seeds are developed. Severe infections result in tremendous yield and quality losses. In addition, infected seeds are contaminated with secondary metabolites produced by these fungi, like the trichothecene mycotoxins, deoxynivalenol and nivalenol. These toxins are hazardous to humans and animals and are causing a variety of acute symptoms (Chelkowski, 1989). Effective fungicides to control the disease are still not available. Therefore, efforts have been made to explore sources of resistance in wheat and rye (Mesterhazy, 1983; Snijders, 1990; Miedaner *et al.*, 1993).

The taxonomy of *Fusarium* spp. has been confused and various classification systems have been proposed (reviewed by Nelson, 1991). Species identification by morphological traits is problematic because characteristics like mycelial pigmentation, formation, shape and size of conidia are unstable and highly dependent on media composition and environmental conditions. Phenotypic variation is abundant and much expertise is required to distinguish between closely related species and to recognize variation within the species (Nelson *et al.*, 1983).

Alternative approaches for identification and differentiation of *F. culmorum* and *F. graminearum* using biochemical (Thrane, 1990) or rRNA sequencing methods (Guadet *et al.*, 1989) did not provide practical systems that are fast, simple, reliable and cost efficient. Recently, Beyer *et al.* (1993) developed an ELISA-based assay for quantitative assessment of *Fusarium* spp. that is specific to various physiological stages of *F. culmorum, F. graminearum* and *F. avenaceum*. However, this ELISA cannot differentiate between the species, indicating their close immunological and, hence, genetic relationship.

A technique for detecting genetic variation has been developed by Welsh and McClelland (1990) and Williams *et al.* (1990), simultaneously, based on the polymerase chain reaction (RAPD-PCR). Genetic variability is assessed by employing short single primer of arbitrary nucleotide sequences. Specific sequence information of the organism under investigation is not required and amplification of genomic DNA is initiated at target sites which are distributed throughout the genome. Polymorphic fragments are the result of variation in the number of appropriate primer-matching sites of different DNAs.

The objectives of our investigations were: (i) to assess the RAPD technique for *Fusarium* spp.; (ii) to develop informative RAPD markers for differentiation between the closely related species *F. culmorum* and *F. graminearum*; and (iii) to determine the genetic variability within and between these two species and to genotype isolates of natural populations.

PROTOCOL

RAPD assay

Materials

1. Laboratory equipment and disposable labware that are routinely used in molecular biology and microbiology. Programmable thermal cycler.
2. *Taq* polymerase supplied with 10 × buffer system, 1 mM dNTP stock solution, decamer oligonucleotide primer kit T from OPERON Technologies, Alameda, California and primer set 100/1 from the University of British Columbia (UBC), Vancouver, Canada. 25 mM MgCl$_2$ stock solution, sterile mineral oil.
3. Standard agarose, 1 × TAE buffer (0.04 mM Tris-acetate, 0.002 mM EDTA, pH 8.0), ethidium bromide (5 μg ml^{-1}), gel loading buffer (Sambrook *et al.*, 1989), chloroform, molecular size standard '100 bp ladder', Pharmacia; Lambda *Bst*EII digested).
4. Crude genomic DNA of fungal isolates.

Fungal isolates were obtained from international culture collections (CBS, Baarn, The Netherlands and DSM, Braunschweig, Germany) and from Dr Th. Miedaner, State Plant Breeding Institute, Stuttgart-Hohenheim, Germany. (Table 7.1 lists the isolates used.)

Likewise, fungal cultures can be prepared from infected plant material according to common isolation procedures (Nelson *et al.*, 1983) with prior surface sterilization and plating of infected plant material on solid media.

Methods

1. Mycelium was harvested from liquid culture grown at 24°C for 4 to 6 days in Petri dishes with 20 ml SNA (special low-nutrient medium; Nirenberg, 1976) supplemented with 0.1% yeast extract.
2. Total genomic DNA was isolated according to the microextraction protocol of Möller *et al.* (1992). In addition, RNA was removed by treating the preparations with 50 μg RNase A (Sambrook *et al.*, 1989) prior to precipitation. DNA was quantified by UV spectrophotometry (260 nm) and agarose gel electrophoresis.
3. PCR was carried out according to the procedure of Williams *et al.* (1990) with minor modifications. Reaction components were set up in bulk mixtures in order to minimize pipetting steps and inaccuracy among samples.

Components	PCR mix for 20 reactions (μl)
10× *Taq* polymerase buffer	50
25 mM MgCl$_2$ stock	20
1 mM stock dNTPs	50
5 μM decamer primer	20
5 U μl^{-1} *Taq* polymerase	4
Sterile distilled water	336
Total volume	480

The mix was partitioned in 24 μl aliquots; 25 ng of genomic DNA in a volume of 1 μl was added to each tube and the reaction was overlaid with 25 μl mineral oil to prevent evaporation.

4. Amplification was performed in a PREM III thermal cycler with the following amplification profile:

Cycle 1	94°C for 3 min
	35°C for 1 min
	72°C for 2 min
Cycles 2–44	94°C for 1 min
	35°C for 1 min
	72°C for 2 min

and a final cycle of 5 min at 72°C.

5. The samples were prepared for electrophoresis by adding 5 μl gel loading buffer. The mineral oil was removed by extraction with 25 μl chloroform.
6. The reaction mix (9 to 15 μl, depending on the size of the gel slot) was loaded and DNA fragments were resolved by electrophoresis (2 V cm^{-1}) in 1.5% agarose and 1 × TAE for 5 h. After staining with ethidium bromide and destaining in distilled water, the gel image was photographed under UV light with a Polaroid camera.
7. Amplification products were visually examined and fragment sizes determined. Presence or absence of each size class were scored as 1 and 0, respectively. The resulting matrix was used to compute Nei's genetic similarity coefficient (Nei and Li, 1979) and UPGMA cluster analysis (Sneath and Sokal, 1973) using the computer package NTSYS-PC written by Rohlf (1989).

Table 7.1. Isolates of *Fusarium* spp. investigated in the presented study.

No.	Isolate	Host	Geographical origin	Source
F. culmorum				
1	FC3.1	Wheat	?	CBS 251.52
2	FC3.7	?	Canada	TM†
3	FC3	Wheat/soil	Norway	TM
4	FC9	Rye	Germany	TM
5	FC11	Wheat	Germany	TM
6	FC16	Rye	Germany	TM
F. graminearum				
7	FG7.1	Wheat	The Netherlands	CBS 389.62
8	FG7.2	Banana	Honduras	CBS 415.86
9	FG7.3	Corn	South Africa	CBS 316.73
10	FG7.5	Corn	Germany	DSM 4528
11	FG7.4*	Wheat	Bulgaria	DSM 62223
12	FG8	Rye	Germany	TM
13	FG10	Rye	Germany	TM
14	FG13*	?	Germany	TM
F. avenaceum				
15	FA3	Rye	Germany	TM
Microdochium nivale				
16	MN	Rye	Germany	TM
Pseudocercosporella herpotrichoides var. *acuformis*				
17	PHA	Rye	Germany	TM

*These isolates were identified to belong truly to the species *F. culmorum.*
† Dr Th. Miedaner, State Plant Breeding Institute, Stuttgart-Hohenheim, Germany.

RESULTS

RAPD assessment and primer screening

The RAPD technique described by Williams *et al.* (1990) was used as starting point for setting up a standard protocol suitable for *Fusarium* species. Primarily, concentration of template DNA, $MgCl_2$, *Taq* polymerase and the thermal cycling profile were varied independently to define reaction conditions which generate reproducible and scorable RAPD profiles. We noticed that template DNA concentrations from 1 to 100 ng had no significant effect on changing the RAPD pattern. Concentrations above 100 ng reduced the number of amplified fragments and with 1 μg of template DNA no products were obtained. This finding was verified with various primers and with template DNA of different isolates. It demonstrates that the amount of template always has to be adjusted for the particular organism under investigation, since size and complexity of its genome seem to have an influence on the PCR performance. Furthermore, varying the annealing temperature from 30 to 38°C had no obvious influence on the RAPD patterns. In contrast, only a narrow range of $MgCl_2$ concentrations (2 to 4 mM) resulted in constantly reproducible patterns. It is important to note that we observed dramatic differences among the commercial sources of *Taq* polymerase. Enzymes of several suppliers did not support amplification to amounts of DNA that are easily detectable by agarose gel electrophoresis. Most likely this is caused by extended thermal stress on the enzyme due to the high number of cycles required for RAPD fragments.

The established protocol was then utilized to generate informative RAPD patterns. One hundred and twenty decamer primers of arbitrary base composition and with a GC-content ranging from 50 to 80% were screened on six single-spore isolates of *F. culmorum* (FC) and of *F. graminearum* (FG) and one isolate each of *F. avenaceum* (FA), *M. nivale* (MN) and *P. herpotrichoides* var. *acuformis* (PHA) from diverse geographical origin. Figure 7.1 demonstrates the high level of profile diversity obtained with ten UBC primers for one isolate of FC and FG each.

RAPD profiles exhibiting distinct fragments for FC and FG, respectively (e.g. Fig. 7.1, lanes 3, 4 and 17, 18) are considered to be informative because these fragments are polymorphic and can be clearly and unambiguously scored. In contrast, faint bands or fragment patterns with high levels of background smear are difficult to score and are prone to be not reproducible.

Out of the 120 primers screened, 88 provided informative RAPD patterns for all isolates, showing one to ten fragments per isolate in the size range of 0.25–3.5 kb.

Species differentiation and genotyping

Thirty-one primers (35%) clearly revealed differences among the three species FC, FG and FA. Intraspecific variability was significantly higher for FG than FC. Profiles of 37

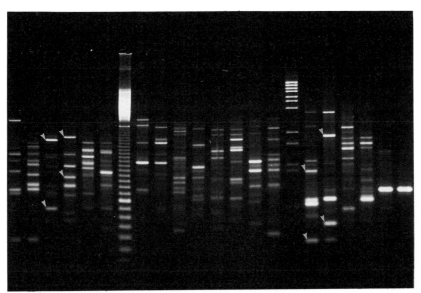

Fig 7.1.   Diversity of fragment patterns obtained with ten different UBC primers of *F. culmorum* (C) and *F. graminearum* (G) isolates. Arrows point to examples of polymorphic fragments. Molecular weight markers are the '100 bp ladder' (M1) of Pharmacia with the 800 bp fragment showing double intensity, and Lambda *Bst*EII digested (M2).

1   2   3   4   5   6 M1 7   8   9 10 12 13 15 M2

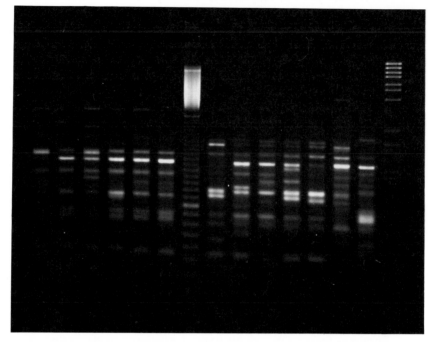

Fig 7.2. RAPD fragments amplified from genomic DNA of different *Fusarium* spp. isolates with primer UBC90 (5'-GGG GGT TAG G-3'). Numbers above the lanes refer to the isolates used. Numbers 1 to 6 represent *F. culmorum* isolates. Numbers 7 to 13 are *F. graminearum* isolates. Number 15 represents a *F. avenaceum* isolate. Size markers (M1 and M2) are as in Fig. 7.1.

were suitable to detect variation among FC isolates. Figs 7.2 and 7.3 show examples of RAPD patterns of different FC, FG, FA, MN and PHA isolates generated with primer UBC90 and UBC85, respectively.

Fragments amplified with UBC90 clearly distinguished between the FC and FG and, in addition, among isolates within each species. Obviously, this primer is likewise suitable for differentiating species and intraspecific genotyping of isolates. Species-specific RAPD patterns were extremely useful to identify false classified isolates, e.g. FG7.4 and FG13 (Fig. 7.3, lane numbers 11 and 14, respectively). In fact, these two isolates belong to the species *F. culmorum*. Various profiles of different primers supported these findings which were also confirmed by careful morphological species determination assisted by *Fusarium* experts.

Data analysis of RAPD profiles obtained with 17 OPERON kit T primers for 13 *Fusarium* spp. isolates and one isolate of *M. nivale* and *P. herpotrichoides* var. *acuformis*, respectively, yielded 178 different fragment classes. The UPGMA cluster analysis clearly grouped FC and FG isolates in two main subclusters (Fig. 7.4) according to their previous species designations. One isolate (FG10) holds an intermediate position, indicating the close relationship of FC and FG. The significant higher intraspecific genetic variability of FG is reflected by a final linkage of only 0.59 versus 0.87 for FC. The associated species FA, MN and PHA are clearly distinct from FC and FG with final very weak linkages of only 0.11. The results of this extensive primer screening demonstrate the usefulness of primers (43%) were highly polymorphic among FG. In contrast, only 12 primers (14%) RAPDs for genotyping and species identification of *Fusarium* spp. and its promising application for analysing the genetic composition of natural fungal populations.

LIMITATIONS

Due to the high sensitivity of PCR, one has to be cautious to avoid contamination in setting up RAPD experiments. Practical guidelines are given by Kwok and Higuchi (1989) and should be considered. 'Good' laboratory practice, well-organized procedures and adhering strictly to a standardized protocol are prerequisites for achieving a high level of reproducibility. However, contamination from traces of another DNA source, e.g. individual fungal spores or bacteria, is generally not troublesome. In fact, PCR of RAPD fragments relies on appropriate priming sites in the target DNA which are more likely to be present in complex genomes. As a consequence, amplification of contaminating DNA from a significantly more complex genome than the investigated one could present a problem. This has been demonstrated with experiments of mixed target DNAs combining genomes of low and high complexity (Williams *et al.*, 1993). In addition, amplification of particular sequences is also a function of the number of available DNA target sites present at the start of the PCR. In most cases, small amounts of such contaminating sequences are considered to have no competing effect and would not be amplified to detectable levels. In every experiment a negative control (PCR reagents omitting target DNA) should be included in order to check the purity (absence of contaminating DNA) of the reaction mix.

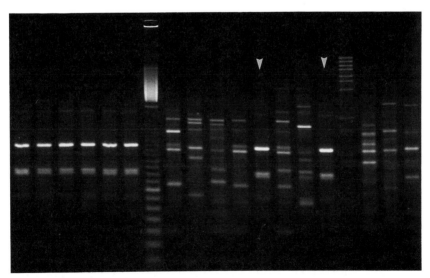

Fig 7.3. RAPD profiles of *Fusarium* spp., *Microdochium nivale* and *Pseudocercosporella herpotrichoides* var. *acuformis* isolates generated with primer UBC85 (5'-GTG CTC GTG C-3'). Lanes 1 to 6 represent *F. culmorum* isolates, lanes 7 to 14 are *F. graminearum* isolates, lanes 15, 16 and 17 represent one isolate of *F. avenaceum*, *M. nivale* and *M. herpotrichoides* var. *acuformis*, respectively. Size markers are as in Fig. 7.2. Patterns marked with an arrow identify these two 'F. graminearum' isolates to be truly *F. culmorum* isolates.

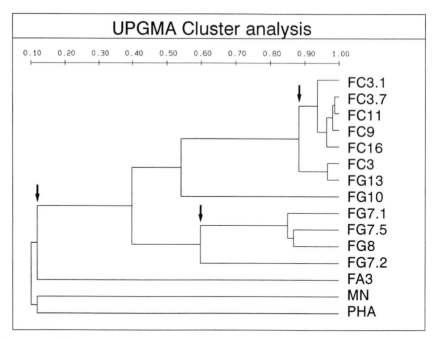

Fig 7.4. UPGMA cluster analysis of RAPD data obtained with 17 OPERON kit T primers for selected *Fusarium* spp., *Microdochium* sp. and *Pseudocercosporella* sp. isolates. Similarity values are indicated and final linkages for the subclusters are marked with arrows.

PERSPECTIVES

To increase specificity of informative RAPDs, we have started to isolate species- and genotype-specific markers by cloning individual fragments. Sequencing and subsequently construction of specific primers will allow us to perform a more robust PCR assay. These so-called 'sequence characterized amplified regions' (SCARs) (Michelmore, 1991) are powerful markers for identification and detection of fungal pathogens in mixed samples and in infected tissue. Furthermore, SCARs will elucidate allelic relationships of RAPD-related markers among isolates and could be used to study their inheritance. The value of this approach has been demonstrated by Paran and Michelmore (1993) with SCARs linked to downy mildew resistance genes in lettuce.

Currently, we are testing more isolates of *Fusarium* spp. with primers that identified informative RAPD patterns in our screening experiment. In addition, we have collected *Fusarium* isolates from naturally infected cereal crops. These will be investigated with characterized RAPD markers and SCARs to obtain an insight into the composition and dynamics of natural pathogen populations and to trace individual isolates in field experiments.

AVAILABILITY OF MATERIAL USED FOR THE ASSAY

Decamer primers of arbitrary sequences are commercially available from OPERON Technologies Inc., Alameda, California, or from the University of British Columbia,

Oligonucleotide Synthesis Laboratory, Vancouver, BC, Canada. The computer package NTSYS-PC is distributed by Exeter Publisher, Setauket, NY, USA.

ACKNOWLEDGEMENT

This work was supported by the Bundesministerium für Forschung und Technologie of Germany (project no. BEO 0318897B).

REFERENCES

Beyer, W., Höxter, H., Miedaner, T., Sander, E. and Geiger, H. H. (1993) Indirect ELISA for quantitative assessment of Fusarium spp. in rye. *Journal of Plant Disease and Protection* (in press).

Chelkowski, J. (ed.) (1989) Formation of mycotoxins produced by Fusaria in heads of wheat, triticale, and rye. In: Chelkowski, J. (ed.) *Fusarium – Mycotoxins, Taxonomy and Pathogenicity.* Elsevier, Amsterdam, pp. 63–85.

Guadet, J., Julien, J., Lafay, J. F. and Brygoo, Y. (1989) Phylogeny of some *Fusarium* species, as determined by large-subunit rRNA sequence comparison. *Molecular Biology and Evolution* 6, 227–242.

Kwok, S. and Higuchi, R. (1989) Avoiding false positives with PCR. *Nature* 339, 237–238.

Mesterhazy, A. (1983) Breeding wheat for resistance to *Fusarium graminearum* and *F. culmorum. Plant Breeding* 91, 295–311.

Michelmore, R. W. (1991) Molecular markers and genome analysis in the manipulation of lettuce downy mildew. Abstract. *Breeding for Disease Resistance.* British Society for Plant Pathology, Newcastle.

Miedaner, T., Borchardt, D. C. and Geiger, H. H. (1993) Genetic analysis of inbred lines and their crosses for resistance to head blight (*F. culmorum, F. graminearum*) in winter rye. *Euphytica* 65, 123–133.

Möller, E. M., Bahnweg, G., Sandermann, H. and Geiger, H. H. (1992) A simple and efficient protocol for isolation of high molecular weight DNA from filamentous fungi, fruit bodies and infected plant tissue. *Nucleic Acids Research* 20, 6115–6116.

Nei, M. and Li, W. H. (1979) Mathematical model for studying genetic variation in terms of restriction endonucleases. *Proceedings of the National Academy of Sciences of the United States of America* 76, 5269–5273.

Nelson, P. E. (1991) History of *Fusarium* systematics. *Phytopathology* 81, 1045–1048.

Nelson, P. E., Toussoun, T. A. and Cook, R. J. (eds) (1981) Fusarium: *Disease, Biology and Taxonomy.* The Pennsylvania State University Press. University Park and London.

Nelson, P. E., Toussoun, T. A. and Marasas, W. F. O. (1983) Fusarium *species. An Illustrated Manual for Identification.* The Pennsylvania State University Press. University Park and London.

Nirenberg, H. I. (1976) Untersuchungen über die morphologische und biologische Differenzierung in der *Fusarium*-Sektion Liseola. *Mitteilungen Biologische Bundesanstalt Land-/ Forstwirtschaft*, Berlin-Dahlem 169, 1–117.

Paran, I. and Michelmore, R. W. (1993) Development of reliable PCR-based markers linked to downy mildew resistance genes in lettuce. *Theoretical and Applied Genetics* 85, 985–993.

Rohlf, F. J. (1989) *NTSYS-pc Numerical Taxonomy and Multivariate Analysis System.* Exeter, New York.

Sambrook, J., Fritsch, E. F. and Maniatis, T. (1989) *Molecular Cloning: A Laboratory Manual*, 2nd edn. Cold Spring Harbor Laboratory Press. Cold Spring Harbor.

Sneath, P. H. A. and Sokal, R. R. (1973) *Numerical Taxonomy*, 2nd edn. Freeman and Company, San Francisco.

Snijder, C. H. A. (1990) The inheritance of resistance to head blight by *Fusarium culmorum* in winter wheat. *Euphytica* 50, 9–17.

Thrane, U. (1990) Grouping *Fusarium* section Discolor isolates by statistical analysis of quantitative high performance liquid chromatographic data on secondary metabolite production. *Journal of Microbiological Methods* 12, 23–39.

Welsh, J. and McClelland, M. (1990) Fingerprinting genomes using PCR with arbitrary primers. *Nucleic Acids Research* 18, 7213–7218.

Williams, J. G. K., Kubelik, A. R., Livak, K. J., Rafalski, J. A. and Tingey, S. V. (1990) DNA polymorphisms amplified by arbitrary primers are useful genetic markers. *Nucleic Acids Research* 18, 6531–6535.

Williams, J. G. K., Hanafey, M. K., Rafalski, J. A. and Tingey, S. V. (1993) Genetic analysis using random amplified polymorphic DNA markers. In: Wu, R. (ed.) *Methods in Enzymology*, vol. 218. Academic Press, San Diego, pp. 704–740.

8

Electrophoretic Detection Method for *Fusarium oxysporum* Species in Cyclamen and Carnation by Using Isoenzymes

A. KERSSIES, A. EVERINK, L. HORNSTRA AND
H. J. VAN TELGEN

Research Station for Floriculture, Aalsmeer, Linnaeuslaan 2a, The Netherlands.

BACKGROUND

Pathogenic fungi can cause great problems in ornamental crops grown under glass. Those that cause vascular diseases and soil fungi usually enter glasshouses on contaminated (young) plants, contaminated culture substrates, men or animals. These fungi, e.g. *Fusarium* and *Phytophthora*, may spread rapidly and contaminate the crop entirely. The cultivation of disease-free plant material under disease-free conditions is vitally important. *Fusarium oxysporum* is a cosmopolitan fungus that exists in many pathogenic forms, parasitizing over 100 species of gymnosperms and angiosperms. *Fusarium oxysporum* f.sp. *cyclaminis* and *F. oxysporum* f.sp. *dianthi* race 2 can cause great problems in the cultivation of *Cyclamen persicum* Mill. and carnation (*Dianthus caryophyllus* L.), respectively. The expression of symptoms is dependent on external circumstances (Rattink, 1986). There is a great need for a sensitive, reliable detection method to test the presence of the fungus in young plants.

An electrophoretic method has been developed for detection of *F. oxysporum* f.spp. in cyclamen and carnation using the isozyme patterns of α-esterase and β-D-glucosidase. The electrophoretic method described is modified for the identification of isolates of *Fusarium oxysporum* f.sp. *dianthi* using isoenzyme patterns of α-esterase (Kerssies, unpublished). Preliminary results indicate that it is possible to distinguish *Fusarium oxysporum* f.sp. *dianthi* from the non-pathogenic *F. oxysporum* (isolated from carnation) and *F. redolens* f.sp. *dianthi* (Everink and Kerssies, 1990; Kerssies and Everink, 1990). Further research on the development of electrophoretic detection methods for other pathogen–host interactions such as *Phytophthora* spp. in potted plants has already started.

Electrophoretic methods for fungi by using isoenzymes

Background to electrophoretic methods for fungi by using isoenzymes

Some studies have been published on the identification of fungi with the use of electrophoresis and isoenzymes. For the identification of *Fusarium* species most of the research is done on the non-specific esterases (Meyer *et al.*, 1964; Meyer and Renard, 1969; Drysale and Bratt, 1971; Matsuyama and Wakimoto, 1977; Szecsi *et al.*, 1976; Ho *et al.*, 1985) and polygalacturonase (Strand *et al.*, 1976; Scala *et al.*, 1981). Other studies on the identification of other fungi by using isoenzymes include Nwaga *et al.* (1990) for *Phytophthora* species; Conti *et al.* (1980) and DaSilva and Azevedo (1983) for *Metarhizium anisopliae* and Powers (1989) for *Cronartium* species.

Very little use has been made of isoenzymes for the detection of fungi in plant material. De Wit *et al.* (1986) showed that at an early stage of the infection process, differences in protein patterns between healthy and diseased plants existed. Hepper *et al.* (1986, 1988a, b) showed that the vesicular–arbuscular mycorrhizal fungi *Glomus caledonium* and *G. mosseae* could be detected in leek roots using isoenzyme patterns of esterase, glutamate oxaloacetate and peptidase.

Electrophoretic detection method for Fusarium oxysporum *species in cyclamen and carnation using isoenzymes*

Cyclamen seedlings were purchased from different commercial growers. Rooted carnation cuttings of 20 different cultivars were used, which varied in their level of resistance to *F. oxysporum* f.sp. *dianthi* race 2 from 0 to 90% resistance.

Cultures of *F. oxysporum* f.sp. *cyclaminis* and *F. oxysporum* f.sp. *dianthi*, isolated from *F. oxysporum*-infected plants were grown on potato dextrose agar (PDA). The inocula of the fungi were prepared by growing the fungi for one week in liquid Czapek Dox medium on a rotary shaker. The cultures were filtered through cheesecloth. Inocula of *F. oxysporum* f.spp. isolated from *Gerbera*, *Aster*, *Chrysanthemum*, *Rosa* and *Pelargonium* were prepared in the same way.

Cyclamen plants were inoculated when they were 14 weeks old by dipping the roots for 30 min in 10 ml fungal suspension (1×10^6 microconidia ml^{-1}). After inoculation the plants were potted in soil. Carnation cuttings were inoculated by pipetting 3 ml conidial suspension (1×10^6 microconidia ml^{-1}) into the potsoil. The inoculum of cultures of *Cylindrocarpon destructans*, *Thielaviopsis basicola*, *Alternaria dianthi* and *Phialophora cinerescens* was prepared by homogenizing cultures grown on PDA or corn meal agar (CMA), and diluted with water to a final concentration of 1×10^6 particles ml^{-1}. The inoculum of cultures of *Erwinia chrysanthemi* and *E. carotovora* was prepared by rinsing the bacteria from a nutrient agar plate with 5–10 ml of 0.15 M phosphate-buffered saline pH 7.2 and with subsequent dilution to a final concentration of 1×10^6 bacteria ml^{-1}. The inoculated and healthy plants were placed on dishes and grown in separated glasshouses at 18°C.

Isoenzyme esterase and glucosidase patterns from extracts of cyclamen and carnation induced by pathogenic, non-pathogenic and other *F. oxysporum* species were compared. These other *F. oxysporum* species were isolated from *Pelargonium*, *Chrysanthemum*, *Aster*, *Gerbera* or *Rosa*. In addition, electrophoresis was carried out with root extracts from cyclamen inoculated with *Cylindrocarpon destructans*, *Thielaviopsis basicola*, *Erwinia chrysanthemi* or *E. carotovora* and with stem extracts from carnations inoculated with *Alternaria dianthi* or *Phialophora cinerescens*.

PROTOCOL

Materials

1. Centrifuge (13,500 rpm).
2. Pestle and mortar.
3. Phast System – Pharmacia.
4.   Extraction buffer for cyclamen: 0.05 M sodium acetate pH 5.0, 1 mM EDTA, 0.04 mM diethyldithiocarbamate (DIECA), 0.34 mM sodium metabisulphite, 0.014% β-mercaptoethanol, 1% polyvinylpyrrolidone (PVP) and 0.5% Triton-X100.

    Extraction buffer for carnation: cyclamen extraction buffer without β-mercapthoethanol
5. Pharmacia ready-to-use-gels:

    PhastGel High Density – α-esterase in cyclamen.

    PhastGel gradient 10–15% – α-D-glucosidase in cyclamen.

    PhastGel gradient 8–25% – α-esterase and β-D-glucosidase in carnation.
6. Pharmacia buffers strips: PhastGel native, pH 8.8.
7. Enzyme staining (for one PhastGel):

    α-esterase

    A 10 mg Fast Blue RR salt in 10 ml 0.15 M phosphate buffer pH 6.9

    B 1 mg 1-Naphthylacetate in 0.2 ml 50% acetone, 50% water.

    Mix A and B, just before using. Incubate in the dark at 30°C for 2 h.

    Rinse with water and fix in 10% acetic acid and 10% glycerol.

    β-D-glucosidase

    A 10 mg Fast Blue BB salt in 10 ml 0.05 mM phosphate buffer pH 6.5

    B 5 mg 6-bromo-2-naphthyl-2-D-glucopyranoside in 0.5 ml acetone

    Mix A and B, just before using. Incubate in the dark at 30°C for 4 hours. Rinse with water and fix in 10% acetic acid and 10% glycerol.
8. Plant material.

Method

Extracts of cyclamen were prepared by grinding 1 g of roots with a pestle and a mortar in liquid nitrogen. The resulting fine powder was then suspended in 1.1 ml cyclamen extraction buffer. The suspension was centrifuged for 30 min at 13,500g (at 4°C) and the supernatant was used for electrophoresis. Extracts of carnation were prepared by cutting 0.5–1.0 g of the stem, from which the leaves were removed, in small pieces and grinding it in the same way as cyclamen roots. The resulting fine powder was suspended in 0.5–1.0 ml carnation extraction buffer and processed as above. Proteins in the samples were separated electrophoretically with the PhastSystem. Aliqouts of each extract (1 μl) were loaded on ready-to-use polyacrylamide gels. Extracts from healthy plants were included on each gel as controls. The type of gel used was dependent on the enzyme activity to be tested.

After electrophoresis enzyme activities were tested for α-D-galactosidase (EC 3.2.1.22), β-D-galactosidase (EC 3.2.1.23), α-esterase (EC 3.1.1.2), β-D-glucosidase (EC 3.2.1.21), leucine amino peptidase (LAP, EC 3.4.11.1), pectin lyase (PL, EC 4.2.99.3), peroxidase (EC 1.11.1.7), glutamate oxaloacetate transaminase (GOT, EC 2.6.1.1) and polygalacturonase (EC 3.2.1.15), according to Tanksley and Orton (1983).

RESULTS AND INTERPRETATIONS

With the enzymes α-D-galactosidase, β-D-galactosidase, LAP, pectin lyase, peroxidase, GOT and polygalacturonase either no activity was found or differences between extracts of healthy and diseased plants were not apparent.

A reproducible difference in isoenzyme patterns between healthy and *Fusarium*-infected cyclamen and carnation plants was found (Fig. 8.1). An extra α-esterase isoenzyme and a β-D-glucosidase enzyme band were formed only in *F. oxysporum*-infected plants. The late appearance of the extra esterase band in carnation stems which was found before, in extracts from cyclamen roots, was probably caused by the use of different plant tissues. The extra esterase isoenzyme band was present in the low molecular part of the gel and that induced by *F. oxysporum* appeared to be identical in cyclamen and carnations. It is not known if this extra esterase isoenzyme was produced by the fungus and essential for the infection process or by the plant as part of the defence mechanism.

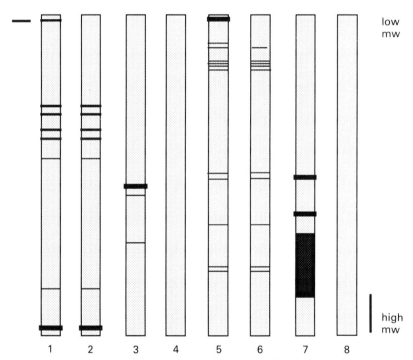

Fig 8.1. Difference in isoenzyme patterns between healthy and *Fusarium*-infected cyclamen and carnation plants.
1. Esterase pattern of cyclamen root extract infected with *Fusarium oxysporum* (—).
2. Esterase pattern of healthy cyclamen root extract.
3. β-D-Glucosidase pattern of cyclamen root extract infected with *F. oxysporum*.
4. β-D-Glucosidase pattern of healthy cyclamen root extract.
5. Esterase pattern of carnation leaf extract infected with *F. oxysporum* (—).
6. Esterase pattern of healthy carnation leaf extract.
7. β-D-Glucosidase pattern of carnation leaf extract infected with *F. oxysporum*.
8. β-D-Glucosidase pattern of healthy carnation leaf extract.

It is remarkable that β-D-glucosidase activity could be detected only in cyclamen and carnation plants infected with *F. oxysporum*. It is known that fungi, such as *Aspergillus niger*, can produce β-glucosidases *in vitro* (Nefisa *et al.*, 1992). Probably *F. oxysporum* produces these glucosidases for metabolization of the sugars present within the plant. The reliability and specificity of the detection method is high. Different genotypes of cyclamen showed little or no variation in the isoenzyme patterns.

The esterase and glucosidase patterns of cyclamen plants did not show the extra enzyme activity observed in *F. oxysporum*-infected plants, when inoculated with *Cylindrocarpon destructans, Thielaviopsis basicola, Erwinia chrysanthemi, E. carotovora*, non-pathogenic *F. oxysporum* or other *F. oxysporum* f.spp. The esterase and glucosidase patterns of carnation plants, also did not show the extra enzyme activity as observed in *F. oxysporum*-infected plants, when inoculated with *Alternaria dianthi, Phialophora cinerescens*, non-pathogenic *F. oxysporum* or other *F. oxysporum* f.spp. If cyclamen plants were inoculated with non-pathogenic or other *F. oxysporum* f.spp., using a very high infection pressure (1×10^6 microconidia ml^{-1}) the extra esterase isoenzyme was sometimes detectable, but not in a reproducible way. Glucosidase activity could not be detected in plants infected with non-pathogenic or other *F. oxysporum* f.spp. Theoretically, pathogenic *F. oxysporum* f.spp. induce symptoms only in specific host plants (Nelson, 1981; Rattink, 1986; Baayen, 1988). Probably the very high inoculation density applied (1×10^6 infectious particles ml^{-1}), evades host–pathogen specificity. Additional experiments showed that a lower infection pressure of the pathogenic *F. oxysporum* delayed the production of the specific isoenzymes.

LIMITATIONS

This method works well under laboratory conditions, but needs more testing under field conditions.

REFERENCES

Baayen, R. P. (1988) *Fusarium* Wilt of Carnation. PhD thesis. Agricultural University of Wageningen, The Netherlands, p. 165.

Conti, E. de, Messias, C. L., Souza H. M. L. de and Azevedo, J. L. (1980) Electrophoretic variation in esterases and phosphatases in eleven wild-type strains of *Metarhizium anisopliae*. *Experientia* 36, 293–294.

DaSilva, P. and Azevedo, J. L. (1983) Esterase pattern of a morphological mutant of the deuteromycete, *Metarhizium anisopliae*. *Transactions of the British Mycological Society* 81, 161–163.

De Wit, P. J. G. M., Buurlage, M. B. and Hammond, K. E. (1986) The occurrence of host-, pathogen- and interaction-specific proteins in the apoplast of *Cladosporium fulvum* (syn. *Fulvia fulva*) infected tomato leaves. *Physiological and Molecular Plant Pathology* 29, 159–172.

Drysale, R. B. and Bratt, P. M. (1971) Electrophoretic patterns of enzymes from isolates of *Fusarium graminearum*. *Transactions of the British Mycological Society* 57, 172–175.

Everink, A. and Kerssies, A. (1990) Development of an electrophoretical detection method for *Fusarium oxysporum* f.sp. *cyclaminis* in *Cyclamen*. In: *Methods for Detecting and Identifying Plant Pathogenic Fungi. Proceedings of the Fourth Cost-88 Workshop*, pp. 45–47.

Hepper, C. M., Sen, R. and Maskall, C. S. (1986) Identification of vesicular–arbuscular mycorrhizal fungi in roots of leek (*Allium porrum* L.) and maize (*Zea mays* L.) on the basis of enzyme mobility during polyacrylamide gel electrophoresis. *New Phytologist* 102, 529–539.

Hepper, C. M., Sen, R., Azion-Aguilar, C. and Grace, C. (1988a) Variation in certain isozymes amongst different geographical isolates of the vesicular–arbuscular mycorrhizal fungi *Glomus clarum, Glomus monosporum* and *Glomus mosseae. Soil Biology and Biochemistry* 20, 51–59.

Hepper, C. M., Azion-Aguilar, C., Rosendahl, S. and Sen, R. (1988b) Competition between three species of *Glomus* used as spatially separated introduced and indigenous mycorrhizal inocula for leek (*Allium porrum* L.). *New Phytologist* 110, 207–215.

Ho, Y. W., Verghese, G. and Taylor, G. S. (1985) Protein and esterase patterns of pathogenic *Fusarium oxysporum* f.sp. *elaeidis* and *F. oxysporum* var. *redolens* from Africa and non-pathogenic *F. oxysporum* from Malaysia. *Phytopathologische Zeitschrift* 114, 301–311.

Kerssies, A. and Everink, A. (1990) Development of an identification technique for *Fusarium oxysporum* f.spp. with electrophoresis. In: *Methods for Detecting and Identifying Plant Pathogenic Fungi. Proceedings of the Fourth Cost-88 Workshop*, pp. 42–44.

Matsuyama, N. and Wakimoto, S. (1977) A comparison of the esterase and calase zymograms of *Fusarium* species with special reference to the classification of a causal fungus of *Fusarium* leaf spot of rice. *Annals of the Phytopathological Society of Japan* 43, 462–470.

Meyer, J. A. and Renard, J. L. (1969) Protein and esterase patterns of two formae speciales of *Fusarium oxysporum. Phytopathology* 59, 1409–1411.

Meyer, J. A., Garber, E. D. and Shaeffer, S. G. (1964) Genetics of phytopathogenic fungi. XII. Detection of esterases and phosphatase in culture filtrates of *Fusarium oxysporum* and *F. xylarioides* by starch-gel zone electrophoresis. *Botanical Gazette* 125, 298–300.

Nefisa, A., Elshayeb, M. A., Mabrouk, S. S., Ismail, S. A. and Abdelfattah, A. F. (1992) Production of fungal enzymes with special reference to β-glucosidases. *Zentralblatt für Mikrobiologie* 147, 563–568.

Nelson, P. E. (1981) Life cycle and epidemiology of *Fusarium oxysporum*. In: Mace, M. E., Bell, A. A. and Beckman, C. H. (eds) *Fungal Wilt Diseases of Plants*. Academic Press, New York, pp. 51–111.

Nwaga, D., Le Normand, M. and Citharel, J. (1990) Identification and differentiation of *Phytophthora* by electrophoresis of mycelial proteins and isoenzymes. *EPPO Bulletin* 20, 35–45.

Powers, H. R. (1989) Interspecific differentiation within the genus *Cronatium* by isozyme and protein pattern analysis. *Plant Disease* 73, 691–694.

Rattink, H. (1986) Some aspects of the etiology and epidemiology of *Fusarium* wilt on *Cyclamen. Mededelingen van de Faculteit Landbouwwetenschappen Rijksuniversiteit Gent* 51/2b, 617–624.

Scala, F., Cristinzio, G., Marziano, F. and Noviello, C. (1981) Endopolygalacturonase zymograms of *Fusarium* species. *Transactions of the British Mycological Society* 77, 587–591.

Strand, L. L., Corden, M. E. and MacDonald, D. L. (1976) Characterization of two polygalacturonase isozymes produced by *Fusarium oxysporum* f.sp. *lycopersici. Biochimica et Biophysica Acta* 429, 870–883.

Szecsi, A., Szentkiralyi, F. and Koves-Pechy, C. (1976) Comparison of esterase patterns of *Fusarium culmorum* and *Fusarium graminearum. Acta Phytopathologica Academiae Scientiarum Hungaricae* 11, 183–203.

Tanksley, S. D. and Orton, T. J. (1983) *Isozymes in Plant Genetics and Breeding, part A*. Elsevier, Amsterdam.

9

Detection and Identification of *Fusarium oxysporum* f.sp. *gladioli* by RFLP and RAPD Analysis

J. J. Mes, J. van Doorn, E. J. A. Roebroeck and P. M. Boonekamp

Bulb Research Centre, PO Box 85, 2160 AB Lisse, The Netherlands.

Background

The fungus *Fusarium oxysporum* Schlecht. is one of the most important and widespread plant pathogens. More than 75 *formae speciales* and over a hundred races can be distinguished by their selective pathogenicity (Snyder and Hansen, 1940; Armstrong and Armstrong, 1981). The nomenclature of isolates is uncertain and controversial. Plant diseases caused by *F. oxysporum* include destructive blights, (vascular) wilts and root rots in important field and vegetable crops such as tomato, pea, melon and potato (Armstrong and Armstrong, 1981). It also reduces the production of many ornamental plants such as carnation and flower bulbs (Nelson *et al.*, 1981).

Fusarium yellows and corm rot of gladiolus and other iridaceous crops is caused by *F. oxysporum* Schlecht.:Fr. f.sp. *gladioli* [(Mass.) Snyder and Hansen] (Massey, 1926; McClellan, 1945; Bald *et al.*, 1971; Boerema and Hamers, 1989). It is considered as a most serious threat and can be devastating in gladiolus grown for flower or corm production. Infection can take place at all stages: during growth in the field and during harvest, cleaning and storage of the corms.

The apparent existence of latent infections in gladiolus corms (Littrell, 1964; Magie, 1966, 1971) causes serious problems in practice. The extent of damage depends on the growing conditions, especially temperature. A certification system for exported corms of 'large-flowered' gladioli to the Mediterranean would be of great economic significance.

A laboratory test based on selective isolation of *Fusarium* from sealed and crushed corms has been developed to detect latent infections (Roebroeck *et al.*, 1990). The method is highly sensitive and shows good correlation with subsequent infection. However, this test cannot distinguish pathogenic from non-pathogenic *Fusarium* spp. A prerequisite for developing a selective detection system is knowledge of the genetic variation of the pathogen. Therefore, we compared host specificity of isolates with genetic relationships between isolates.

Two races of *Fusarium oxysporum* f.sp. *gladioli* could be identified by pathogenicity tests on two differential gladiolus cultivars: race 1 is pathogenic to both 'large- and small-flowered' gladiolus cultivars and race 2 is pathogenic only to 'small-flowered' cultivars (Roebroeck and Mes, 1992). Furthermore, we identified four vegetative compatibility groups (VCG 0340, 0341, 0342 and 0343) among 57 isolates of *F. oxysporum* f.sp. *gladioli*, based on complementation tests with nitrate-non-utilizing mutants. Within VCG 0340 both races existed (Roebroeck and Mes, 1992; Mes *et al*, 1993). For routine diagnosis, race 1 (pathogenic to 'large-flowered' gladioli) is the most important group of isolates to be detected.

Protocols

RFLP analysis

1. Isolate DNA using the cetyl trimethyl ammonium bromide (CTAB) method described by Manicom *et al.* (1987), with omission of the second extraction step.
2. Digest total genomic DNA with *Hind*III according to the manufacturer's instructions (Boehringer Mannheim GmbH, Germany).
3. Separate restriction digests on 1% agarose gel.
4. Blot DNA (PosiBlotTM pressure blotter) from gel to Hybond N$^+$ membrane (Amersham International plc, Amersham, UK).
5. Hybridize blots with horseradish-peroxidase-labelled probe D4 and incubate according to standard instructions (Amersham, ECL-kit).
6. Wash blots and expose them to Kodak X-ray films.
7. Assign a position number to each band with a different electrophoretic mobility.
8. Transform the RFLP pattern to a binary code by giving each band the value '1' when it is present at a certain position or the value '0' when it is absent.
9. Calculate similarity according to Digby (1984) with the help of a computer program (Payne *et al.*, 1987).

RAPD analysis

1. Isolate DNA according to the rapid extraction procedure described by Cenis (1992).
2. Perform PCR in volumes of $50 \mu l$ containing 10 mM Tris-HCl, pH 9.0, 1.5 mM MgCl$_2$, 50 mM KCl, 0.01% gelatine, 1% Triton X-100, $100 \mu M$ each of dATP, dCTP, dGTP and dTTP (Perkin Elmer Cetus), $0.2 \mu M$ primer, 25 ng of DNA and 1 unit of *Taq* DNA polymerase (HT Biotechnology Limited).
3. Overlay mineral oil to prevent evaporation.
4. Amplify in a thermocycler for 30–40 cycles after initial denaturation for 4 min at 94°C. Each cycle consists of 1 min at 94°C, 2 min at 36°C and 2 min at 72°C. The last cycle has a final extension at 72°C for 10 min.
5. Resolve the amplification products by electrophoresis in a 2% agarose gel and stain with ethidium bromide.
6. Blot PCR products and digested total fungi DNA from gel to Hybond N$^+$.
7. Isolate the discriminating band from the gel using the Geneclean II kit (Bio 101, Inc., USA) and reamplify the band if necessary, using the same primer.

8. Hybridize blots with labelled probe, according to the protocol described above (see RFLP analysis, steps 5 and 6).

RESULTS AND LIMITATIONS

RFLP analysis of *Fusarium oxysporum* f.sp. *gladioli*

Restriction fragment length polymorphism analysis was performed to determine the genetic relationship between isolates. RFLPs were detected by Southern hybridization analysis in which a random DNA fragment of total DNA of an isolate of *F. oxysporum* f.sp. *dianthi* (Manicom *et al.*, 1987) was used as a probe. Only a part of this 3400 bp probe (D4) was responsible for most of the multibanding patterns. This 376 bp fragment has been studied in detail but its function is still unknown (Manicom *et al.*, 1990). The probe was chosen because vegetative compatibility groups (VCGs) within *F. oxysporum* f.sp. *dianthi* could be identified, and homology with *F. oxysporum* f.sp. *gladioli* could be revealed (Manicom *et al.*, 1987, 1990).

We have analysed 57 isolates, collected from widely separated geographic regions, for the existence of races, VCG and RFLP patterns.

Pathogenic isolates could be divided into three RFLP groups (Table 9.1, RFLP groups I, II and III) and two remaining isolates showing single distinctive RFLP patterns (IV and V) (Mes *et al*, 1993). Non-pathogenic isolates exhibited different patterns (VI–XVII) or because they showed no homology with the probe. The RFLP groups corresponded with VCGs, not with races. Isolates within VCG 0340 belonging to either races 1 or 2 cannot be distinguished by this probe. This makes the method unsuitable for routine diagnosis. The RFLP patterns of isolates were found to be similar or identical within VCGs, but quite different between VCGs. We conclude from these results that the *forma specialis gladioli* probably has a polyphyletic origin. We suggest that a VCG/RFLPG is a group which has obtained pathogenicity for a host plant or a group of host plants independently from other VCG/RFLPGs. Therefore, development of races from different VCGs can be based on different mechanisms. By differential screening of races 1 and 2 within VCG0340/RFLPGI small differences could be revealed, which could be used as specific probes for detection of one race only.

Table 9.1. Correlation between races, VCGs and RFLP analysis of *Fusarium* isolates obtained from gladiolus corms.

Number of Isolates	Race	VCG	RFLP
29	1	0340	I
3	2	0340	I
6	2	0341	II
2	2	034–*	IV
1	1	0343	V
12†	—	—	VI–XVII
3†	—	—	—

* Isolate not capable of forming a heterokaryon.
† Non-pathogenic isolates.

RAPD analysis of *Fusarium oxysporum* f.sp. *gladioli*

The method we are using now for development of a test for race 1 isolates of RFLP group I is the PCR-based RAPD method (Williams *et al.*, 1990). Comparing banding patterns obtained after PCR with random 10-mer oligonucleotide primers (Operon Technologies, Inc., Alameda, CA), should identify primers that are specific for race 1 isolates of RFLP group I.

We selected four isolates: two race 1/RFLP I isolates (G2 and Ir7) and two race 2/RFLP I isolates (G6 and Ir1). Genomic DNA samples from these genetically closely

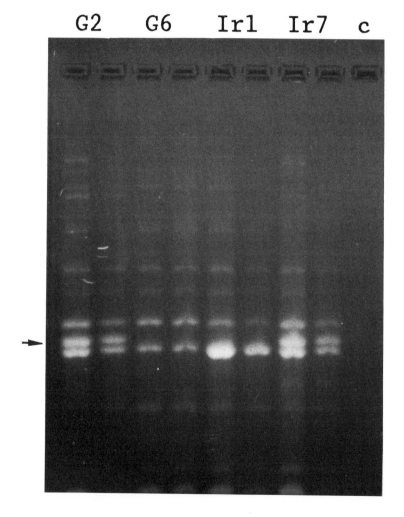

Fig 9.1. Patterns of amplification products of four isolates (two DNA concentrations) after PCR with OP A-01 primer and electrophoresis.
G2 and Ir7 are race 1 isolates, G6 and Ir1 are race 2 isolates. The control (c) is the same reaction but without DNA. The arrow marks the discriminating 850 bp fragment.

related isolates were used in amplification reactions, performed in a Bio-med Thermo-cycler 60. The PCR protocol was essentially as described by Williams *et al.* (1990).

So far, we have tried 80 different primers (OP A–D, 20 primers per set). With 23 of these primers we observed polymorphism between isolates. Only one primer (OP A-01) gave a band which distinguished race 1 isolates from race 2 isolates (Fig. 9.1).

This discriminating band, present in G2 and Ir7 but absent in the PCR product of G6 and Ir1, was isolated from gel (Geneclean II kit, Bio 101, Inc., USA), amplified again, purified and used as a probe after it was checked for contaminating products. Specificity of this band was analysed by Southern hybridization experiments. A blot with PCR products of a broader range of isolates amplified with primer A-01 as well as a blot with *Hind*III digested total genomic DNA of these isolates were both hybridized with the labelled probe. Unfortunately, the discriminating band was not as specific as was hoped for. The band was present in all isolates (both in PCR products and in total *Fusarium* DNA), but in much lower concentrations in race 2 isolates. This can be explained by either assuming lower copy numbers in the genome of race 2 or base pair changes in the primer binding side and other parts of the genomic sequence to which the probe hybridizes. Testing shorter stretches of the probe or analysing the DNA sequence has to elucidate whether or not this fragment can be used.

REFERENCES

Armstrong, G. M. and Armstrong, J. K. (1981) *Formae speciales* and races of *Fusarium oxysporum* causing wilt diseases. In: Nelson, P. E., Toussoun, T. A. and Cook, R. J. (eds) *Fusarium: Diseases, Biology, and Taxonomy.* Pennsylvania State University Press, University Park, pp. 391–399.

Bald, J. G., Suzuki, T. and Doyle, A. (1971) Pathogenicity of *Fusarium oxysporum* to Easter lily, Narcissus and Gladiolus. *Annals of Applied Biology* 67, 331–342.

Boerema, G. H. and Hamers, M. E. C. (1989) Check-list for scientific names of common parasitic fungi. Series 3b: Fungi on bulbs: Amaryllidaceae and Iridaceae. *Netherlands Journal Plant Pathology* 95 (Suppl. 3), 1–32.

Cenis, J. L. (1992) Rapid extraction of fungal DNA for PCR amplification. *Nucleic Acids Research* 20, 2380.

Digby, P. G. N. (1984) Dendrograms and ziggurats. *Genstat Newsletter* 14, 14–18.

Littrell, R. H. (1964) Studies on latent *Fusarium* in gladiolus corms. *Annual Report Florida Agricultural Experiment Stations* pp. 315–316.

Magie, R. O. (1966) Gladiolus *Fusarium* disease development and control. *Gladiolus* 41, 106–110.

Magie, R. O. (1971) Carbon dioxide treatment of gladiolus corms reveals latent *Fusarium* infections. *Plant Disease Reporter* 55, 340–341.

Manicom, B. Q., Bar-Joseph, M., Rosner, A., Vigodsky-Haas, H. and Kotze, J. M. (1987) Potential application of random DNA probes and restriction fragment length polymorphisms in the taxonomy of the Fusaria. *Phytopathology* 77, 669–672.

Manicom, B. Q., Bar-Joseph, M., Kotze, J. M. and Becker, M. M. (1990) A restriction fragment length polymorphism probe relating vegetative compatibility groups and pathogenicity in *Fusarium oxysporum* f.sp. *dianthi. Phytopathology* 80, 336–339.

Massey, L. M. (1926) *Fusarium* rot of gladiolus corms. *Phytopathology* 16, 509–523.

McClellan, W. D. (1945) Pathogenicity of the vascular *Fusarium* of gladiolus to some additional iridaceous plants. *Phytopathology* 35, 921–930.

Mes, J. J., van Doorn, J., Roebroeck, E. J. A., van Egmond, E., van Aartrijk, J. and Boonekamp, P. M. (1993) Restriction fragment length polymorphisms, races and vegetative compatibility groups within a worldwide collection of *Fusarium oxysporum* f.sp. *gladioli*. *Plant Pathology* (in press).

Nelson, P. E., Horst, R. K. and Woltz, S. S. (1981) Fusarium diseases of ornamental plants. In: Nelson, P. E., Toussoun, T. A. and Cook, R. J. (eds) *Fusarium: Diseases, Biology, and Taxonomy*. Pennsylvania State University Press, University Park, pp. 121–128.

Payne, R. W., Lane, P. W., Ainsley, A. E., Bicknell, K. E., Digby, P. G. N., Harding, S. A., Leech, P. K., Simpson, H. R., Todd, A. D., Verrier, P. J., White, R. P., Gower, J. C., Tunnincliffe Wilson, G. and Paterson, L. J. (1987) *Genstat 5 Reference Manual*. Clarendon Press, Oxford.

Roebroeck, E. J. A. and Mes, J. J. (1992) Physiological races and vegetative compatibility groups within *Fusarium oxysporum* f.sp. *gladioli*. *Netherlands Journal of Plant Pathology* 98, 57–64.

Roebroeck, E. J. A., Groen, N. P. A. and Mes, J. J. (1990) Detection of latent *Fusarium oxysporum* in gladiolus corms. *Acta Horticulturae* 266, 469–476.

Snyder, W. C. and Hansen, H. N. (1940) The species concept in *Fusarium*. *American Journal of Botany* 27, 64–67.

Williams, J. G. K., Kubelik, A. R., Livak, K. J., Rafalski, J. A. and Tingey, S. V. (1990) DNA polymorphisms amplified by arbitrary primers are useful as genetic markers. *Nucleic Acids Research* 18, 6531–6535.

10 RFLPs for Distinguishing Races of the Pea Vascular Wilt Pathogen *Fusarium oxysporum* f.sp. *pisi*

A. Coddington,[1] B. G. Lewis[1] and D. S. Whitehead[2]

[1] *School of Biological Sciences, University of East Anglia, Norwich NR4 7TJ, UK;* [2] *Department of Plant Pathology, 120 Long Hall, Clemson University, Clemson, South Carolina 29631, USA.*

Background

Fusarium oxysporum Schlecht emend. Snyd. and Hans. f.sp. *pisi* (Van Hall) Snyd. and Hans. which causes *Fusarium* wilt of peas, was one of the most damaging diseases of the crop until control measures were introduced; it remains a major threat if cultivar resistance sources become ineffective or if control is relaxed.

Although long-range dissemination on contaminated seeds has been reported, the disease is mainly soil-borne and inoculum survives in contaminated soil for many years. The pathogen infects the fibrous roots of pea, advancing into the vascular tissue of susceptible cultivars. Initial symptoms include curling of the leaves and petioles; subsequently, leaves become chlorotic in a progressive sequence from the older to the young ones. At high temperatures (>20°C) the disease progresses rapidly and most of the foliage on diseased plants becomes desiccated and dies. Median longitudinal sections of the infected roots and stem base reveal an orange or reddish pigmentation of the vascular tissue.

On the basis of standardized pathogenicity tests on seven differential host lines (Table 10.1), four genetic variants of the pathogen, called races 1, 2, 5 and 6, have been distinguished (Kraft and Haglund, 1978). Races 1 and 2 have a widespread distribution in many pea-growing regions of the world whereas 5 and 6 were thought to be localized in western USA. Recently, isolates with race 6 characteristics on the seven differential lines have been reported in Europe (Boedker *et al.*, 1992). Macroscopic symptoms caused by these four races on susceptible host plants are indistinguishable (Haglund and Kraft, 1970; Kraft *et al.*, 1974, 1981) but on a field scale, race 2 causes a less severe disease (near wilt) than the other races. There are also differences in cultural characteristics between race 2 and the others. On potato dextrose agar at pH 4.5, races 1, 5 and 6 show little pigmentation whereas race 2 typically produces a purplish-black pigment. Only race 2 sporulates profusely in culture, producing abundant macrospores and microspores; the other races produce few microconidia and usually no macroconidia.

Table 10.1. Classification of race isolates of *Fusarium oxysporum* f.sp. *pisi* by their pathotypes on the seven host differentials.

	Races			
Cultivar	1	2	5	6
Little Marvel	S	S	S	S
Darkskin Perfection	R	S	S	S
New Era	R	R	S	S
New Season	R	R	S	R
WSU 23	R	R	R	S
WSU 28	R	S	R	R
WSU 31	R	R	R	R

Source: The table is adapted from Coddington *et al.* (1987) which gives full details of the pathogenicity test used.
R = resistant; S = susceptible.

Control of these diseases by the use of resistant cultivars has been highly successful. Resistance to race 1, based on the F_w gene, has proved remarkably durable (Lewis and Matthews, 1985) and is present in about half of the currently used commercial cultivars of pea. Sources of resistance, also thought to be based on single dominant genes, are available for each of the other races (Hare *et al.*, 1949; Kraft *et al.*, 1981) However, strategies for deploying these sources in cultivars and the use of resistant cultivars in disease control, require reliable tests for distinguishing races. In practice, variation in culture and pathogenicity causes problems in distinguishing the races. The pathogenicity test on differential lines needs very careful standardization of inoculation procedures and environmental control; variation between testing centres is well known (Coddington *et al.*, 1987).

PROTOCOL

To obtain a race-specific RFLP requires the following steps:

1. Production of fungal mycelium.
2. DNA extraction.
3. Restriction digestion.
4. Agarose gel electrophoresis.
5. Transfer of DNA to a nylon membrane (Southern transfer).
6. Preparation of probe DNA.
7. Hybridization and detection.

Steps 1 to 4 are described in detail in Coddington and Gould (1992).

Southern transfer or 'blotting'

Materials

A nylon membrane, such as Hybond-N$^+$ (Amersham International) is used. The latter is particularly good if the ECL non-radioactive probing and detection system is used.

Method

For subsequent use of radioactive probes capillary transfer under neutral conditions is used (Sambrook *et al.*, 1989). When nucleic acid probes are to be labelled and detected by the non-radioactive ECL system the transfer procedure is that recommended by the manufacturer of that system (Amersham International).

Preparation of probe DNA

Materials

Probes were derived from the plasmid pDG106 whose construction is described in Whitehead *et al.* (1992).

pDG106 contains a moderately repetitive 3kb DNA fragment from *Fusarium oxysporum* f.sp. *pisi* race 1 cloned into the *Hind*III site of pUC18.

Method

The insert DNA is cut out from the plasmid using *Hind*III, separated from the vector DNA on an agarose gel and extracted from the gel using a GeneClean kit. (Stratech Scientific). Radioactive labelling was achieved using $[a\text{-}^{32}P]dCTP$ with an Amersham Multiprime labelling kit (Amersham International) and following the manufacturer's protocols. Non-radioactive labelling was achieved using the ECL system (Amersham International), again following the manufacturer's instructions.

Hybridization and detection

For the radioactively labelled probes, prehybridization, hybridization and posthybridization washes followed the Amersham protocols. Following overnight hybridization at 65°C, membranes were washed in 2×SSC (standard saline citrate) for 15 min at 65°C. After drying they were then exposed to X-ray film at −80°C for 24 h. For the probes labelled using the non-radioactive enhanced chemiluminescence (ECL) system, prehybridization, hybridization and posthybridization conditions were according to the Amersham protocols.

RESULTS

RFLP patterns for representative isolates of the four common races using the insert from pDG106 as probe and the ECL detection method, are shown in Fig. 10.1 (taken from Whitehead *et al.*, 1992). Races 1 and 5 each have a characteristic pattern which is the same for all isolates tested. Race 2 isolates, on the other hand, had one of two patterns (A and B) suggesting that race 2 comprises two subgroups which are not distinguished by the standard host differentials. The one available race 6 isolate tested had the same pattern as race 1 and hence cannot be distinguished from race 1 by this method. Analogous RFLP patterns, using the same insert from pDG106 radioactively labelled, are shown in

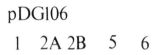

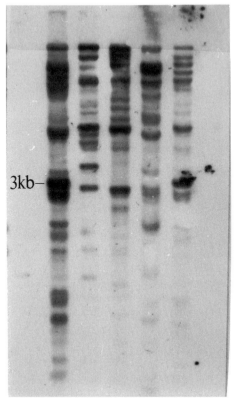

Fig 10.1   Southern blots of *Fusarium* DNA cut with *Hind*III and probed with insert DNA from pDG106 labelled using the non-radioactive ECL technique. Lanes 1 to 5: isolates of races 1, 2A, 2B, 5 and 6, respectively.

Fig. 10.2 (taken from Boedker *et al.*, 1993). Again the patterns of races 1 and 6 are the same and race 2 isolates had one of two patterns.

A new finding from this work was that three newly isolated race 6 strains from Denmark had the same RFLP pattern but this was different from that of the previously characterized isolates of races 1 and 6 (Fig. 10.2). This was the first report of a race 6 outside the United States and it was surprising to find that it had a different RFLP pattern to the American one. Isolates of all the other races had the same pattern irrespective of geographical origin.

Hence RFLP analysis can distinguish easily between isolates of existing races and also detect the presence of new groupings which have not previously been defined as separate races by the use of existing host differentials. The technique is quicker and more reliable than pathogenicity testing and there is usually very little variation in RFLP pattern among isolates of the same race or race subgrouping.

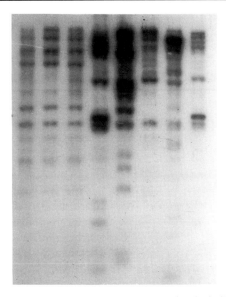

Fig 10.2   Southern blots of *Fusarium* DNA cut with *Hind*III and probed with radioactively labelled insert DNA from pDG106. Lanes 1 to 3: three new race 6 isolates from Denmark. Lanes 4 to 8: isolates of races 1, 2A, 2B, 5 and 6, respectively.

LIMITATIONS

Assigning a new isolate to a race by its RFLP pattern is only possible if the correlation between the pattern and an existing race is already established. For example, the newly isolated race 6 strains from Denmark would not have been assigned to race 6 on the basis of the patterns shown in Fig. 10.2. Other limitations to the method are the costs of both the radioactive and non-radioactive labelling and detection systems and the need to have special facilities for the radioactive work. Hence it will not be possible to adapt it easily to a field situation and especially so to one in developing countries.

AVAILABILITY OF THE TEST

The plasmid pDG106 can be obtained from the authors as plasmid DNA in *Escherichia coli* strain JM83.

REFERENCES

Boedker, L., Lewis, B. G., Coddington, A. and Smedegaard, V. (1992) Discovery in Europe and characteristics of *Fusarium oxysporum* f.sp *pisi* race 6. *Proceedings of the 1st European Conference of Grain Legumes*, Angers, pp. 367–368.

Boedker, L., Lewis, B. G. and Coddington, A. (1993) The occurrence of a new genetic variant of *Fusarium oxysporum* f.sp. *pisi*. *Plant Pathology* (in press).

Coddington, A. and Gould, D. S. (1992) Use of RFLPs to identify races of fungal pathogens. In:

Duncan, J. M. and Torrance, L. (eds) *Techniques for the Rapid Detection of Plant Pathogens.* Blackwell Scientific, Oxford.

Coddington, A., Matthews, P. M., Cullis, C. and Smith, K. H. (1987) Restriction digest patterns of total DNA from different races of *Fusarium oxysporum* f.sp. *pisi* – an improved method for race classification. *Journal of Phytopathology* 118, 9–20.

Haglund, J. M. and Kraft, W. A. (1970) *Fusarium oxysporum* f.sp. *pisi* race 5 *Phytopathology* 60, 1861–1862.

Hare, W. W., Walker, J. G. and Delwich, E. J. (1949) Inheritance of a gene for near wilt resistance in the garden pea. *Journal of Agricultural Research* 78, 239–250.

Kraft, J. M. and Haglund, W. A. (1978) A reappraisal of the race classification of *Fusarium oxysporum* f.sp. *pisi. Phytopathology* 68, 273–275.

Kraft, J. M., Muehlbauer, F. J., Cook, R. J. and Entemann, F. (1974) The reappearance of common wilt of peas in eastern Washington. *Plant Disease Reporter* 58, 62–64.

Kraft, J. M., Burke, D. W. and Haglund, W. A. (1981) *Fusarium* diseases of beans, peas and lentils. In: Nelson, P. E., Toussoun, T. A. and Cook, R. J. (eds) *Fusarium: Diseases, Biology and Taxonomy.* Pennsylvania State University Press, University Park, USA and London.

Lewis, B. G. and Matthews, P. (1985) The world germplasm of *Pisum sativum*; could it be used more effectively to produce healthy crops? In: Hebblethwaite, P. D., Heath, M. C. and Dawkins, T. C. K. (eds) *The Pea Crop – a Basis for Improvement.* Butterworths, London.

Sambrook, J., Fritsch, E. F. and Maniatis, T. (1989) *Molecular Cloning: A Laboratory Manual*, 2nd edn. Cold Spring Harbor, New York.

Whitehead, D. S., Coddington, A. and Lewis, B. G. (1992) Classification of races by DNA polymorphism analysis and vegetative compatibility grouping in *Fusarium oxysporum* f.sp *pisi. Physiological and Molecular Plant Pathology* 41, 295–305.

11 The Use of Amplified Fragment Length Polymorphisms (AFLPs) of Genomic DNA for the Characterization of Isolates of *Fusarium oxysporum* f.sp. *ciceris*

A. Kelly,[1,2] A. R. Alcalá-Jiménez,[3] B. W. Bainbridge,[2] J. B. Heale,[1] E. Pérez-Artés[3] and R. M. Jiménez-Díaz[3]

[1] *Plant Cell and Molecular Sciences Group, and* [2] *Microbial Physiology Group, King's College, University of London, Campden Hill Road, London W8 7AH, UK;* [3] *Instituto de Agricultura Sostenible, CSIC, Apartado 3048, 14080 Cordoba, Spain.*

Background

Fusarium wilt, induced by *Fusarium oxysporum* Schlecht: Fr. f.sp. *ciceris* (Padwick) Matuo and K. Sato, is one of the most important soil-borne diseases of chickpea (*Cicer arietinum* L.) throughout the world and particularly in the Indian subcontinent and the Mediterranean basin (Haware, 1990; Nene and Reddy, 1987).

Symptoms of the disease can develop within 25 days after sowing or up to podding stage, depending upon susceptibility of cultivars. Early wilting causes more loss than late wilting; seeds from late-wilted plants are lighter, rougher and duller than those from healthy plants (Haware and Nene, 1980). The disease syndrome can also vary with isolates of the pathogen. Isolates of *F. oxysporum* f.sp. *ciceris* from southern Spain induce either yellowing or wilt syndromes as a result of vascular infections (Trapero-Casas and Jiménez-Díaz, 1985). The yellowing syndrome is characterized by a progressive foliar yellowing which develops 30–40 days after inoculation and late death of the plant. The wilt syndrome develops 15–20 days after inoculation and is characterized by chlorosis, flaccidity and early plant death.

Isolates of *F. oxysporum* f.sp. *ciceris* also differ in pathogenicity to chickpea lines and can be classified into pathogenic races. Haware and Nene (1982) first identified races 1,2,3 and 4 in India. Later, three additional races, namely races 0,5, and 6 were identified in southern Spain (Jiménez Díaz *et al.*, 1989). Additional work with a larger number of *F. oxysporum* f.sp. *ciceris* isolates from California, Morocco, northern and southern Spain and Tunisia, indicated that slight phenotypic differences occur among isolates within a race and therefore they could be classified into pathogenicity or race groups, namely race groups 0,1,5 and 6 (Jiménez Díaz *et al.*, 1993). Race group 0 is the least virulent of all race groups identified. Isolates of this race group induce the yellowing syndrome. All other isolates induce wilt, except for some race group 1 isolates that induce yellowing (Jiménez Díaz *et al.*, 1993).

Pathogenicity tests to characterize disease syndrome and race group among isolates of *F. oxysporum* f.sp. *ciceris* are expensive, time consuming, and may be influenced by variability inherent in the experimental system (Bhatti and Kraft, 1992; Alcalá-Jiménez *et al.*, 1992). Thus, it would be desirable to have a rapid, cheaper and less labour intensive method.

Restriction fragment length and amplified fragment length polymorphisms of fungal DNA

Restriction fragment length polymorphisms (RFLPs) utilizing mitochondrial, ribosomal or total DNA have proved very useful in a wide variety of fungi (Coddington *et al.*, 1987; Forster *et al.*, 1990; Typas *et al.*, 1992). In fungi, the mitochondrial DNA (mtDNA) is small and present in high copy number (Taylor, 1986), which makes it suitable for restriction enzyme analysis. Analysis of mtDNA from *F. oxysporum* f.sp. indicates that significant variability in mtDNA restriction fragment patterns occurs within some *formae speciales*, i.e. *F. oxysporum* f.sp. *niveum* (E. F. Sm.) W. C. Snyder and H. N. Hans. (Kim *et al.*, 1992), *F. oxysporum* f.sp. *melonis* W. C. Snyder and H. N. Hans. (Jacobson and Gordon, 1990); but no variation was found in mtDNA within others, i.e. *F. oxysporum* f.sp. *conglutinans* (Wollenweb.), W. C. Snyder and H. N. Hans., *F. oxysporum* f.sp. *mathioli* K. Baker, and *F. oxysporum* f.sp. *raphani* J. B. Kendrick and W. C. Snyder (Kistler *et al.*, 1987; Kistler and Benny, 1989).

The recent development of random amplification of polymorphic DNA (RAPD) (Williams *et al.*, 1991) or arbitrarily primed polymerase chain reaction (AP-PCR) (Welsh and McClelland, 1991) has allowed the generation of reliable, reproducible DNA fragments or fingerprints in several fungi (Crowhurst *et al.*, 1991; Goodwin and Annis, 1991; Guthrie *et al.*, 1992; Kush *et al.*, 1992). These techniques utilize short single primers of arbitrary or random base sequence, with over 50% G+C content to amplify genomic DNA at low stringency conditions. Only the sequences that have proximal priming sites in the correct orientation will be amplified and due to the low stringency, some mismatch annealing may occur giving rise to further products. The purpose of our study was to use RFLPs of mtDNA and amplified fragment length polymorphisms (AFLPs) of total DNA to distinguish between isolates of different race groups of *F. oxysporum* f.sp. *ciceris*, and to use this information to provide a rapid method for diagnosis of race type as well as to predict pathogenicity of unknown isolates. Here we report the method used and results obtained with AFLPs.

Protocol

AFLPs of total DNA

Materials

1. Petri plates, Erlenmeyer flasks.

2. PDA, PDB.
3. Miracloth filter.
4. Buchner funnel.
5. Mortar and pestle.
6. Sterile Corex centrifuge tubes (15 ml), 4 mm glass beads, polypropylene and polycarbonate centrifuge tubes.
7. Lysis solution (10 mM Tris-HCl pH 8; 250 mM EDTA pH 8; proteinase K (Sigma) 200 μg ml^{-1}; Triton-X100, 0.5% v/v).
8. 1.5 M NaCl, 1.5 mM MgCl$_2$, phenol, chloroform, isopropanol.
9. RNase A (Sigma, heated to inactivate DNAase).
10. TE buffer (10 mM Tris-HCl pH 7.6, 1 mM EDTA pH 8).
11. Primers, deoxynucleotide triphosphates, *Taq* DNA polymerase and reaction buffer (Promega).
12. Agarose (Sigma).
13. Tris-acetate buffer (40 mM Tris-acetate, 1 mM EDTA).
14. Ethidium bromide.
15. Phage lambda DNA.
16. *Hind*III restriction enzyme.
17. Test material (single spore cultures of *F. oxysporum* f.sp. *ciceris*).

Methods

1. Maintain stock cultures of *F. oxysporum* f.sp. *ciceris* on PDA in Petri plates at 25°C and 12 h photoperiod of fluorescent and near-UV light at approx 36 μE m^{-2} s^{-1}.
2. Grow cultures in PDB at 110–120 rpm for 3 days at same temperature and light conditions as above.
3. Harvest some 15–20 g of fresh mycelium by filtration through Miracloth in a Buchner funnel.
4. Freeze mycelial mats in liquid nitrogen and lyophilize them.
5. Grind up lyophilized mycelium (500 mg) in a pestle and mortar and transfer the powder to a sterile 15 ml Corex centrifuge tube.
6. Add six sterile 4 mm glass beads to tubes, cap them with Parafilm and vortex for 2–15 min with cooling in ice until mycelium is a fine, uniform powder.
7. Add 5–7.5 ml of lysis solution and mix the powder by inversion.
8. Transfer the mixture to a sterile polypropylene centrifuge tube with a screw cap, and incubate overnight at 37°C with gentle shaking.
9. Add 0.15 vol. of 1.5 M NaCl followed by 0.7 vol. of phenol and 0.3 vol. of chloroform, and mix the tube by inversion.
10. Centrifuge for 1 h at 20,000g at room temperature.
11. Remove the supernatant carefully leaving behind the protein interface, and transfer it to a polypropylene centrifuge tube.
12. Add 0.05 vol. of RNase A, incubate at 37°C for 10 min, add 1 vol. of chloroform and mix by inversion for 5 min.
13. Centrifuge at 13,000g for 10 min and remove the supernatant. Record the volume and transfer it into a polycarbonate centrifuge tube.
14. Add 0.54 vol. of isopropanol dropwise and invert the tube quickly to mix. Tap the tube gently until DNA precipitates as a clot (few minutes).

15. Decant the supernatant carefully.
16. Wash the pellet of DNA with 70% alcohol, dry under vacuum and redissolve in 50 μl of TE buffer.
17. Perform reactions (100 μl: 05 μM of primer, 200 μM of deoxynucleotide triphosphates, 10 μl of 10 × reaction buffer, 1.5 mM $MgCl_2$ and 100 ng of fungal DNA) on a thermal cycler, with the following steps.
18. Denaturation at 94°C for 5 min followed by a 'hold' step at 72°C to allow addition of *Taq* DNA polymerase.
19. Allow 30 cycles of 1 min annealing temperature (40–45°C depending upon primers), 3 min extension at 72°C and 1 min denaturation at 94°C.
20. Allow a final cycle of 1 min of annealing followed by 6 min at 72°C.
21. Analyse 20 μl volume of each reaction on a 2% agarose gel run at 1.5 V cm^{-1} overnight in 1× Tris-acetate buffer and stain with ethidium bromide.
22. Digest phage lambda DNA with *Hind*III and use restriction fragments as size markers.

RESULTS

AFLPs of total DNA

DNA extracted from 73 fungal isolates was analysesd for AFLP patterns using three primers (Table 11.1). These isolates comprised 63 *F. oxysporum* f.sp. *ciceris* isolates from California (7), India (8) and Spain (48); five *F. oxysporum* f.sp. *melonis* isolates from Cyprus (4) and Spain (1); two *F. oxysporum* f.sp. *niveum* isolates from Cyprus (1) and Spain (1); and one each of *F. eumartii*, *F. solani* and *Ascochyta rabiei* from Spain. Isolates from Cyprus were supplied by Dr C. A. Pallos, Dept. of Agriculture, Nicosia, Cyprus. Of the 63 isolates of *F. oxysporum* f.sp. *ciceris*, 42 had been characterized to race previously and represent all race groups and geographical origins (Jiménez-Díaz *et al.*, 1993). However, pathogenicity information was withheld for 20 of them which were used in a 'blind trial'.

    Amplification with primer P2 gave rise to a band of 1.3 kb (Fig. 11.1) in all isolates of *F. oxysporum* f.sp. *ciceris* (except isolate Fo-8250) which cause the yellowing syndrome,

Table 11.1. List of primers used in this study, together with their percentage GC content, annealing temperature and origin.

Primer	Origin*	Sequence 5′–3′	G + C (%)	Annealing temperature (°C)
P2	*P. hordei* rDNA	CACCGCCCCAAAATGGCCAC	65	45
P6	*P. hordei* rDNA	GTCCTCAGTCCCCCAATCCC	65	45
KS	pBluescript™	CGAGGTCGACGGTATCG	65	40

* The *Penicillium hordei* primers were originally derived from sequence analysis of the IGS of the ribosomal RNA gene complex. The KS primer is a synthetic sequencing primer used in sequencing pBluescript recombinants.

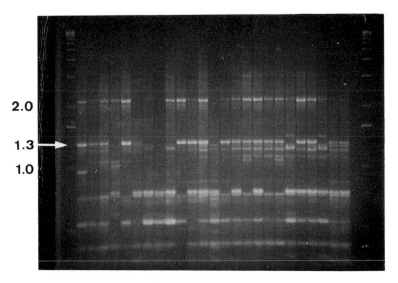

Fig 11.1.   AFLP fingerprint of *Fusarium oxysporum* f.sp. *ciceris* generated by primer P2. The outer lanes are 1 kb DNA markers (Gibco BRL). Numbers on the side are sizes in kilobase pairs. The arrow indicates the 1.3 kb band found only in isolates that induce the yellowing syndrome.

irrespective of their race identity. In addition, a band of 0.55 kb was found to be present in all *F. oxysporum* f.sp. isolates, but absent in other chickpea pathogens such as *A. rabiei*, *F. eumartii* and *F. solani*. With primer KS, two bands of approximately 1.6 kb were observed in wilt-inducing *F. oxysporum* f.sp. *ciceris* isolates but not in the yellowing isolates

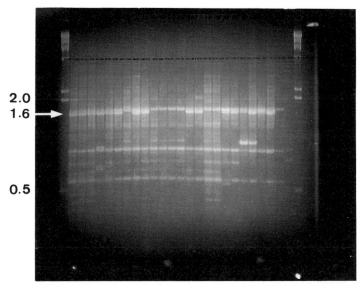

Fig 11.2.   AFLP fingerprints of *Fusarium oxysporum* f.sp. *ciceris* generated by primer KS. The outer lanes are *Hind*III fragments of phage lambda DNA. Numbers on the side are sizes in kilobase pairs. The arrow indicates the 1.6 kb band found only in wilt-inducing isolates.

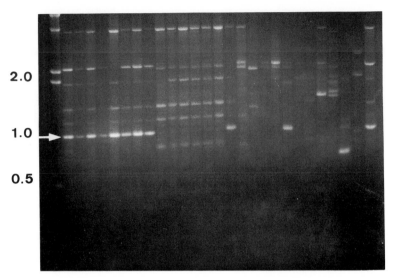

Fig 11.3. AFLP fingerprints of *Fusarium oxysporum* f.sp. *ciceris* (lanes 2–15) generated by primer P6. The outher lanes are *Hind*II fragments of phage lambda DNA. Numbers on the side are sizes in kilobase pairs. Lanes 16–26 correspond to *F. oxysporum melonis, F. oxysporum niveum, F. eumartii, F. solani* and *Ascochyta rabiei.* The arrow indicates the 1.0 kb band found only in *F. oxysporum ciceris* isolates that induce the yellowing syndrome.

(Fig. 11.2). A 1.0 kb band in yellowing isolates and race 3 isolates was also observed. Primer 6 generated bands of 2.4 kb and 1.0 kb in the yellowing isolates (Fig. 11.3), and a separable 2.3 kb band in wilt-inducing isolates. No bands specific to geographical origin were observed.

The 22 isolates of *F. oxysporum* f.sp. *ciceris* used in a 'blind trial' were correctly typed as to pathogenicity, i.e. the disease syndrome that they induce. Furthermore, 21 *F. oxysporum* f.sp. *ciceris* isolates of unknown pathogenicity, which have similar AFLP profiles to yellowing (13) or wilting (8) isolates, were subsequently shown in inoculation of race differentials to induce the corresponding disease syndrome (Alcalá-Jiménez *et al.*, unpublished).

AFLP data from single primer PCR were used to construct a data matrix from which a dendrogram could be generated using unweighted pair grouping with arithmetic averaging (UPGMA) (Sneath and Sokal, 1973). UPGMA analysis of the AFLPs separated the isolates into three distinct clusters. All the yellowing isolates (except Fo-8250) fall into one cluster with 6% dissimilarity. Isolate Fo-8250 is 12% dissimilar from other race group 0 isolates. All the wilt-inducing isolates form another cluster with 6% dissimilarity between them (except isolate Fo-8012 which is 8% dissimilar). Ten other isolates used in this study, comprising isolates of *A. rabiei, F. eumartii, F. oxysporum* f.sp. *melonis, F. oxysporum* f.sp. *niveum,* and *F. solani* form a loose, third group, 22% dissimilar from *F. oxysporum ciceris.*

CONCLUSIONS

Using single primers in a PCR-based reaction, it was possible to separate isolates of *F. oxysporum* f.sp. *ciceris* into one of two groups. These groupings correlate well with the disease syndrome that they induce in chickpea. It was also possible to place previously unknown isolates into one of the two groups, the reliability of this approach having been demonstrated in the 'blind trial'.

F. oxysporum f.sp. *ciceris* was distinguished from other *formae speciales* and from other chickpea pathogens such as *A. rabiei, F. eumartii* and *F. solani*, using AFLPs. An analysis which can clearly distinguish between fungi likely to be present in the same host or environment is obviously an advantage.

The above findings suggest that AFLP analysis provides a rapid method of differentiating between isolates of *F. oxysporum* f.sp. *ciceris* inducing chickpea yellowing or wilt. Further work is needed to discriminate between isolates of the pathogen belonging to different race groups. Current work involves further AFLP analysis using other primers and the characterization of the syndrome-specific bands, to develop race group specific primers or probes for the *in planta* or in soil detection and quantification of *F. oxysporum* f.sp. *ciceris*.

Acknowledgements

Research supported in part by the Commission of the European Communities ECLAIR programme, Contract AGRE 0051; and Comisión Interministerial de Ciencia y Tecnología (CICYT), grant AGF92–0910-C02–01.

References

Alcalá-Jiménez, A. R., Trapero-Casas, J. L. and Jiménez Díaz, R. M. (1992) Influencia de la densidad de inóculo en la infección de garbanzo por *Fusarium oxysporum* f.sp. *ciceris*. In: Romero Muños, F. and Gómez Barcina, A. (eds), *Resúmenes del VI Congreso Latino-Americano de Fitopatología*. Consejería de Agricultura y Pesca, Junta de Andalucía, Sevilla, Spain, 39 pp.

Bhatti, M. A. and Kraft, J. M. (1992) The effects of inoculum density and temperature on root rot and wilt of chickpea. *Plant Disease* 76, 50–54.

Coddington, A., Matthews, P. M., Cullis, C. and Smith, K. H. (1987) Restriction digest patterns of total DNA from different races of *Fusarium oxysporum* f.sp. *pisi* an improved method for race classification. *Journal of Phytopathology* 118, 9–20.

Crowhurst, R. N., Hawthorne, B. T., Rikkerink, E. H. A. and Templeton, M. D. (1991) Differentiation of *Fusarium solani* f.sp. *cucurbitae* races 1 and 2 by random amplification of polymorphic DNA. *Current Genetics* 20, 391–396.

Forster, H., Oudemans, P. and Coffey, M. D. (1990) Mitochondria and nuclear DNA diversity within six species of *Phytophthora*. *Experimental Mycology* 14, 18–31.

Goodwin, P. H. and Annis, S. L. (1991) Rapid identification of genetic variation and pathotype of *Leptosphaeria maculans* by randomly amplified polymorphic DNA assay. *Applied and Environmental Microbiology* 57, 2482–2486.

Guthrie, P. A. I., Magill, C. W., Frederiksen, R. A. and Odvody, G. N. (1992) Randomly amplified polymorphic DNA markers: a system for identifying and differentiating isolates of *Colletotrichum graminicola*. *Phytopathology* 82, 832–835.

Haware, M. P. (1990) Fusarium wilt and other important diseases of chickpea in the Mediterranean area. *Options Méditerranéennes Série Séminaires* 9, 61–64.

Haware, M. P. and Nene, Y. L. (1980) Influence of wilt at different growth stages on the yield loss in chickpea. *Tropical Grain Legume Bulletin* 19, 38–44.

Haware, M. P. and Nene, Y. L. (1982) Races of *Fusarium oxysporum* f.sp. *ciceri*. *Plant Disease* 66, 809–810.

Jacobson, D. J. and Gordon, T. R. (1990) Variability of mitochondrial DNA as an indicator of relationships between populations of *Fusarium oxysporum* f.sp. *melonis*. *Mycological Research* 94, 734–744.

Jiménez-Díaz, R. M., Trapero-Casas, A. and Cabrera de la Colina, J. (1989) Races of *Fusarium oxysporum* f.sp. *ciceri* infecting chickpeas in southern Spain. In: Tjamos, E. C. and Beckman, C. H. (eds) *Vascular Wilt Diseases of Plants*, Vol. H28. Springer-Verlag, Berlin, pp. 515–520.

Jiménez-Díaz, R. M., Alcalá-Jiménez, A. R., Hervás, A. and Trapero-Casas, J. L. (1993) Pathogenic variability and host resistance in the *Fusarium oxysporum* f.sp. *ciceris/ Cicer arietinum* pathosystem. *Proceedings of the 3rd European Seminar on Fusarium Mycotoxins, Taxonomy, Pathogenicity and Host Resistance*. Hodowla Roślin Aklimatyzacja Inasiennichwo (special edition) 37, 1–4 and 87–94.

Kim, D. H., Martyn, R. D. and Magill, C. W. (1992) Restriction fragment length polymorphism groups and physical map of mitochondrial DNA from *Fusarium oxysporum* f.sp. *niveum*. *Phytopathology* 82, 346–353.

Kistler, H. C. and Benny, U. (1989) The mitochondrial genome of *Fusarium oxysporum*. *Plasmid* 22, 86–89.

Kistler, H. C., Bosland, P. W., Benny, U., Leong, S. and Williams, P. H. (1987) Relatedness of strains of *Fusarium oxysporum* from crucifers measured by examination of mitochondrial and ribosomal DNA. *Phytopathology* 77, 1289–1293.

Kush, R. S., Becker, E. and Wach, M. (1992) DNA amplification polymorphisms of the cultivated mushroom *Agaricus bisporus*. *Applied and Environmental Microbiology* 58, 2971–2977.

Nene, Y. L. and Reddy, M. V. (1987) Chickpea diseases and their control. In: Saxena, M. C. and Singh, K. B. (eds) *The Chickpea*. CAB International, Wallingford, England, pp. 233–270.

Sneath, P. and Sokal, R. (1973) *Numerical Taxonomy: The Principles and Practices of Numerical Classification*. Freeman, San Francisco.

Taylor, J. W. (1986) Fungal evolutionary biology and mitochondrial DNA. *Experimental Mycology* 10, 259–269.

Trapero-Casas, A. and Jiménez-Díaz, R. M. (1985) Fungal wilt and root rot of chickpeas in southern Spain. *Phytopathology* 75, 1146–1151.

Typas, M. A., Griffen, A. M., Bainbridge, B. W. and Heale, J. B. (1992) Restriction fragment length polymorphisms in mitochondrial DNA and ribosomal RNA gene complexes as an aid to the characterization of species and subspecies populations in the genus *Verticillium*. *FEMS Microbiology Letters* 95, 157–162.

Welsh, J. and McClelland, M. (1991) Fingerprinting genomes using PCR with arbitrary primers. *Nucleic Acids Research* 18, 7213–7218.

Williams, P. H., Kubelik, A. R., Livak, K. J., Rafolski, J. A. and Tingey, S. V. (1991) DNA polymorphisms amplified by arbitrary primers are useful as genetic markers. *Nucleic Acids Research* 18, 6531–6535.

12 PCR-based Assays for the Detection and Quantification of *Verticillium* Species in Potato

J. ROBB,[1] X. HU,[1] H. PLATT[2] AND R. NAZAR[1]

[1] *Department of Molecular Biology and Genetics, University of Guelph, Guelph, Ontario, Canada N1G 2W1;* [2] *Agriculture Canada Research Station, Charlottetown, Prince Edward Island, Canada.*

BACKGROUND

Fungi of the genus *Verticillium* cause vascular wilt disease in many different crop plants and are an important factor in most agricultural economies throughout the world (Pegg, 1981). *V. albo-atrum* and *V. dahliae* are the most widespread and most destructive species; infection results in a variety of symptoms which can include vascular browning, stunting, leaf flaccidity, foliar chlorosis, leaf abscission, reduced yield and ultimately death. Unfortunately, disease incidence and severity has been increasing steadily during the last fifty years (Pegg, 1981). No doubt, a major factor which has contributed to the spread of *Verticillium*, has been our inability to identify and quantify adequately these pathogen species. Current detection methods usually involve the plating of plant parts or washings (e.g. from seed) on selective medium (Pegg and Street, 1984). This diagnostic test is at best semi-quantitative and, in the case of seed-borne infection, can detect inoculum only on the seed surface.

Both *V. dahliae* and *V. albo-atrum* are important and somewhat different pathogens of potato (Busch, 1967; Pegg, 1974). The resultant wilt disorder severely reduces yield and quality. In addition, this host can be infected with a third less pathogenic *Verticillium* species, *V. tricorpus* (Isaac, 1952; Smith, 1965) which frequently is associated with the potato early dying complex. Producers suffer further losses in seed sales and usage as infected plants produce tubers which are infected and/or infested with these three pathogens. Colonization of the tubers also causes an increase in soil-borne levels of inoculum, and hence, an ever-increasing disease potential in the fields used for potato production. *Verticillium* wilt results from the planting of infested seed or planting seed in contaminated soil. Susceptible weed species, the presence of symptomless carriers and soil-borne population levels maintained by non-hosts are additional important contributors (Pegg, 1974).

Constant and rapid monitoring of pathogen levels in potato is very important in regions where *Verticillium* is a major disease. Potato cultivars can differ widely in levels of resistance to various *Verticillium* species (Platt, 1986) and climate and/or cultivar changes can lead to critical shifts in the occurrence and/or predominance of *V. dahliae* and *V. albo-atrum* (Celetti and Platt, 1987). Such observations emphasize the need for an accurate method of species identification and quantification which is sufficiently rapid for a producer or inspector to make appropriate recommendations since control options differ depending on pathogen type. At present, the main method of detection is field inspection; however, *Verticillium* wilt can have the same plant symptoms as excess chemical top-killer, drought stress and mature plant senescence. To determine the actual causal agent and frequency of occurrence, plants and soils must be taken to laboratories for assessment. Unfortunately, the methods available for pathogen assay are very time consuming (up to 7 weeks), labour intensive and often of limited accuracy. Results are at best semi-quantitative and most current testing techniques fail to distinguish *V. albo-atrum* from *V. dahliae* and other similar species.

The development of a molecular probe which could be used for rapid (hours to days) identification and quantification of *Verticillium* species would clearly have a major impact on various sectors of the potato industry as well as a commercial value. Several attempts have been made to develop an ELISA test for this purpose, so far without success (Lazarovitz, 1987). Again, a major stumbling block has been the failure to distinguish between *V. albo-atrum* and *V. dahliae*. Recent studies in our laboratories (Nazar *et al.*, 1991; Hu *et al.*, 1993) have indicated the feasibility of developing molecular probes based on genomic DNA sequencing, recombinant DNA techniques and newly developed PCR technology (Mullis *et al.*, 1986).

Background to PCR assays for *Verticillium* spp.

Microbial detection assays based on PCR technology have numerous advantages over traditional approaches (Henson and French, 1993). Identification can be made from cultures or *in planta* with equal facility and very small quantities of pathogen can be detected with or without the use of radioactivity. PCR is rapid and versatile; also, if tests are properly designed, several assays for different pathogens can be carried out simultaneously on the same sample.

Our PCR-based assays for the detection of *Verticillium* have utilized small differences in the structure of the ribosomal genes for the design of primer sets capable of differentiating among species. Because of their ease of isolation and relatively high gene copy number, ribosomal genes have been used frequently for the development of molecular probes and PCR-based assays (see Nazar *et al.*, 1991; Henson and French, 1993). While the nucleotide sequences of mature rRNAs are highly conserved (Robb *et al.*, 1993), both the non-transcribed and transcribed spacer sequences are more divergent. For example, comparing the intravening transcribed spacer (ITS) regions, which are approximately 295 bp long in *Verticillium*, five nucleotide differences were observed between *V. dahliae* and *V. albo-atrum* (Nazar *et al.*, 1991), 17 between *V. dahliae* and *V. tricorpus* (Robb *et al.*, 1993) and 12 between *V. albo-atrum* and *V. tricorpus*. These differences were used to develop the differential primer sets used in the assays for *V. dahliae* (Nazar *et al.*, 1991), *V. albo-atrum* (Nazar *et al.*, 1991) and *V. tricorpus* (Moukhamedov *et al.*, 1993) which are described in this chapter.

Recent studies in our laboratories (Robb *et al.*, 1993) have shown that there are two distinct subgroups of *V. albo-atrum* which can be found in potato. *V.a.a./*1 responds positively with the original *V. albo-atrum* primers (Nazar *et al.*, 1991) but *V.a.a./*2 could not be identified with any existing assay. When the ITS regions of the *V.a.a./*2 ribosomal genes were sequenced, the ITS regions were found to be more similar to those of *V. tricorpus*. A unique primer set (Robb *et al.*, 1993) was designed for the differentiation of *V.a.a./*2 (Fig. 12.1) and screening of *V. albo-atrum* potato isolates from North America and Europe revealed that *V.a.a./*2 can be found in the United Kingdom, the Netherlands and Canada. The distribution and biological basis of the *V.a.a./*2 subgroup is currently under further investigation.

PROTOCOL: PCR-BASED ASSAYS FOR DETECTION OF *VERTICILLIUM* SPP. IN POTATO

These are general-purpose assays developed for the detection of *V. dahliae*, *V. albo-atrum* (*V.a.a./*1 and *V.a.a./*2) and *V. tricorpus* in a wide range of plant species including *Solanum tuberosum* (Nazar *et al.*, 1991; Hu *et al.*, 1993; Moukhamedov *et al.*, 1993; Robb *et al.*, 1993). The tests can be used with or without radioactivity. They can be applied effectively to analysis of isolated fungal cultures, roots, tubers, stems, leaves, seed and soil. We have also provided a protocol which includes the addition of an internal standard (Fig. 12.2) for quantitative analyses of fungal biomass in any of the above milieux.

Preparation of plant and fungal DNA

Fungal cultures were grown and maintained on potato dextrose agar, V8 juice agar or selective media. To obtain mycelia or conidia, fungal colonies were grown in Czapek's broth (Tuite, 1969) at 22°C in the dark with shaking. Spores were harvested ten days after inoculation by filtration through two layers of cheesecloth.

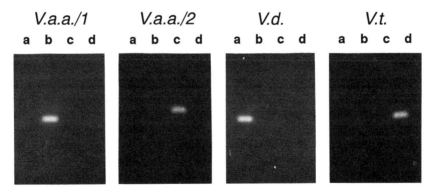

Fig 12.1.   Detection and differentiation of *Verticillium* spp. by a standardized polymerase chain reaction assay. DNA extracted from fungal spores of potato isolates of *V. dahliae* (lane a), *V. albo-atrum*/1 (lane b) and /2 (land c), and *V. tricorpus* (lane d) were PCR amplified in 1 mM Tris-HCl, pH 8.3, 50 mM KCl, 1.5 mM MgCl$_2$, 0.2 mM BSA, 0.2 mM dNTP using 12.5 pmol of *V. albo-atrum*/1 (*V.a.a./*1), *V. albo-atrum* (*V.a.a./*2), *V. dahliae* (*V.d.*), or *V. tricorpus* (*V.t.*)-specific primers and 2 units of Taq DNA polymerase. The PCR reaction consisted of 30 cycles of denaturation at 95°C, annealing at 60°C and polymerization at 72°C.

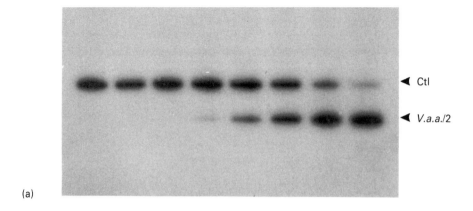

(a)

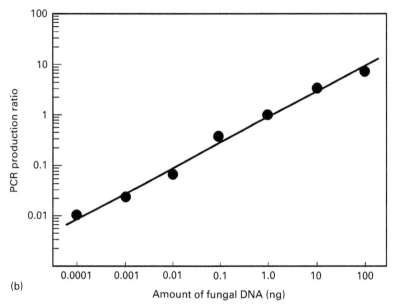

(b)

Fig 12.2.   Relationship between the amount of fungal DNA and the resulting PCR product ratio. Varying amounts (0.0001–100 ng) of *V.a.a.*/2 genomic DNA were PCR amplified in the presence of 0.1 pg of heterologous internal control template (Ctl). The resulting products were fractionated on an 8% polyacrylamide gel, detected by autoradiography (a) and quantified by liquid scintillation counting (b). In the autoradiograph, the first lane on the left contained only control template DNA; the remaining lanes (left to right) correspond to the seven points in (b). The regression line in (b) was determined as $y = ax^b$ from the PCR product ratios (*V. tricorpus* product/internal control product).

Plant tissues were cut into small pieces in order to facilitate the extraction of DNA. For quantitative analyses of fungal biomass in plant tissues the plant material must be weighed before extraction.

Genomic DNA from *Verticillium* and potato was extracted by one of two methods: the hexadecyl trimethylammonium bromide (CTAB) method of Rogers and Bendich (1985) as modified by Nazar *et al.* (1991), or an SDS-phenol method (Steele *et al.*, 1965).

When a CTAB buffer was used, filtered *Verticillium* hyphae or spores, or finely-cut plant pieces were ground to a coarse powder in the presence of liquid nitrogen; 1–3 g of ground tissue were suspended in extraction buffer (1.4 M NaCl, 20 mM EDTA, 0.1 M Tris-HCl, pH 8.0 containing 1% PVP-40 and 2% CTAB). The suspension was then extracted twice with chloroform/isoamyl alcohol and precipitated with 2 volumes of ethanol containing 2% potassium acetate. When SDS buffer was used, approximately 0.8 g of plant material was cut into small pieces (1 mm lengths) and ground to a coarse powder in the presence of liquid nitrogen and resuspended in 4 ml of 0.3% SDS, 140 mM NaCl, 50 mM sodium acetate (pH 5.1); 300 μl of homogenate were mixed with 300 μl phenol saturated with TE buffer, 10 mM Tris-HCl, pH 8.0, 1 mM Na EDTA, for 10 min. The phases were separated by centrifugation and the aqueous layer was further extracted with an equal volume of chloroform/isoamyl alcohol (24:1, v/v). The nucleic acid was precipitated from the aqueous phase with 2 volumes of ethanol containing 2% potassium acetate, washed once with 95% ethanol and dissolved in 20 μl of TE buffer.

Preparation of internal control templates for quantitative analyses

Control DNA templates were prepared using *Verticillium* spp.-specific primer sequence borders and a heterologous internal sequence (Hu *et al.*, 1993). The standard species-specific primers were used to amplify genomic DNA from an unrelated fungus, *Fusarium oxysporum*, under non-specific annealing conditions (37°C for 2 min). These reactions yielded multiple PCR products. Fragments were chosen to be larger (*ca.* 500 bp for *V. tricorpus*, and *V.a.a.*/2) or smaller (200 bp for *V. dahliae*, and *V.a.a.*/1) than the PCR amplified fungal fragments. The selected fragments were eluted after fractionation on polyacrylamide gels, cloned into pTZR19 (Mead *et al.*, 1986), and verified using the dideoxy sequencing methods of Sanger *et al.*, (1977) as modified for direct plasmid sequencing by Zhang *et al.*,(1988). The development of the internal controls for *V. dahliae* and *V. albo-atrum*/1 have been described by Hu *et al.*, (1993) and the control for *V. tricorpus* assays by Moukhamedov *et al.*, (1993). Figure 12.2 illustrates the use of the new *V. albo-atrum*/2 standards.

The internal control plasmid template DNAs were prepared using the method of Holmes and Quigley (1981) and all DNA standards were purified further by CsCl density gradient centrifugation (Radloff *et al.*, 1967). The amount of DNA was determined from the absorbancy at 260 nm or by comparison with a known standard after fractionation by agarose gel electrophoresis (Maniatis *et al.*, 1982; McDonnell *et al.*, 1977).

PCR assay protocol

Materials and equipment

1. 10 × PCR buffer – 500 M KCl, 100 mM Tris-HCl (pH 9.0 at 25°C), 15 mM $MgCl_2$, 0.1% gelatin (w/v), and 1% Triton X-100.
2. Nucleotide stock solution – 2 mM of each of dATP, dCTP, dTTP and dGTP in autoclaved deionized water.
3. *Taq* DNA polymerase – Perkin-Elmer (Norwalk, CT, USA) Ampli-Taq[R] DNA polymerase or the equivalent.

4. Primers were synthesized using a Cyclone 'Plus' automated oligonucleotide syn-thesizer (Milligen/Biosearch, Milford, MA, USA).
5. Temperature cycler – Perkin-Elmer (Norwalk, CT, USA) DNA Thermal Cycler or the equivalent.
6. Mini-gel apparatus – Pharmacia Model GNA-100 (Sweden) or the equivalent.
7. Loading dye solution – formamide containing 0.1% bromophenol blue and 0.1% xylene cyanol.
8. Staining solution – 0.5 μg ml^{-1} ethidium bromide in deionized water.

Method

1. DNA extract is used in quantities of 0.01–0.05 μg per reaction tube.
2. PCR amplification is conducted using a 50 μl aliquot of PCR reaction mixture containing 1× PCR buffer, 0.2 mM bovine serum albumin (BSA), 0.2 mM of each deoxyribonucleotide triphosphate, 12.5 pmol of each oligonucleotide primer, 2 units of *Taq* DNA polymerase and the sample DNA. Variations: (a) for quantitative studies also add 0.1 pg of each control plasmid DNA; (b) for radioactive studies also add 0.5 μCi[a-^{32}P]dATP per reaction tube.
3. Add 60 μl of light mineral oil and vortex twice for 5 s each. Separate the phases by centrifugation in a microfuge for 1 min.
4. Using a temperature cycler, amplify the DNA with 30 reaction cycles consisting of a 1 min denaturation step at 95°C, a 1 min annealing step at 60°C and a 2 min elongation step at 72°C.

(a) For qualitative analyses

5. Heat 5 μl of PCR reaction mixture with an equal volume of loading dye to 65°C. Fractionate on a 1% agarose gel (Maniatis *et al.*, 1982). Carry out electrophoresis at 80–90 V for 2 h using a mini-gel apparatus.
6. Stain gel for 20 min with ethidium bromide staining solution.
7. Visualize bands with a UV transilluminator (Fig. 12.1).

(b) For quantitative analyses

8. Heat 5 μl of PCR reaction mixture with an equal volume of loading dye solution to 65°C and fractionate on an 8% polyacrylamide gel sequencing gel (Maniatis *et al.*, 1982). Carry out electrophoresis at 600 V for 3 h.
9. Carry out autoradiography using fast Kodak X-ray film with an exposure of 2 h at 70°C (Fig.12.2a). Orient the autoradiograph to the gel plate; mark PCR products and cut out the relevant bands.
10. Quantify the signals using a liquid scintillation counter.
11. Calculate the ratio of intensity of control signal to the fungal signal.
12. Determine amount of fungal DNA present from the standard curve (Fig. 12.2b).

LIMITATIONS

Some plant tissues contain natural inhibitors of the PCR reaction (e.g. Hu *et al.*, 1993). Inhibitory effects can readily be detected by a decrease in the staining intensity of the internal control. For this reason, whenever plant extracts are used, we always include a control reaction using only the internal control as substrate. The inhibitory effect can frequently be relieved by suitable dilution of the extracts with water before PCR analysis. The levels of inhibition tend to vary from tissue to tissue and plant to plant and must be determined empirically for each sample. In potato, a dilution of 1/10 has been found to be sufficient for most stem extractions but dilutions of 1/100 or greater may be required for some tubers. Presently, we are investigating alternative methods of chemically removing the inhibitor(s) from the extract.

With or without dilution, false negatives may be obtained in quantitative analyses if the level of pathogen in the tissues is very low. In such cases, the use of an internal control template may outcompete the test sample. If a negative result is obtained in any reaction including an internal control, one should repeat the test omitting the control template. A positive signal without the presence of the control template indicates trace amounts of the pathogen below levels quantifiable by the assay.

AVAILABILITY OF PRIMERS AND INTERNAL CONTROLS

The DNA sequences for the ITS1 and 2 regions of *V. dahliae*, *V. albo-atrum*/1 and 2 and *V. tricorpus* are available from the EMBL Data Library.

REFERENCES

Busch, L. V. (1967) Distribution in Ontario of *Verticillium* strains causing wilt of potatoes. *Canadian Plant Disease Survey* 47, 76–78.

Celetti, M. J. and Platt, H. W. (1987) A new cause for an old disease: *Verticillium dahliae* found on Prince Edward Island. *American Potato Journal* 64, 209–212.

Henson, J. M. and French, R. (1993) The polymerase chain reaction and plant disease diagnosis. *Annual Reviews of Phytopathology* 31 (in press).

Holmes, D. S. and Quigley, M. (1981) A rapid method of preparation of bacterial plasmids. *Analytical Biochemistry* 114, 193–194.

Hu, X., Nazar, R. N. and Robb, J. (1993) Quantification of *Verticillium* biomass in wilt disease development. *Physiological and Molecular Plant Pathology* (in press).

Isaac, I. (1952) A further comparative study of pathogenic isolates of *Verticillium*: *V. nubilum* Pethybr. and *V. tricorpus* sp. *Transactions of the British Mycological Society* 36, 80–195.

Lazarovitz, G. (1987) Detection of *Verticillium* antigens by dot-ELISA on nitrocellulose membranes. *Canadian Journal of Plant Pathology* 9, 82.

Maniatis, T., Fritsch, E. F. and Sambrook, J. (1982) *Molecular Cloning: A Laboratory Manual*. Cold Spring Harbor Laboratory, Cold Spring Harbor, NY, pp. 309–365.

McDonnell, M. W., Simon, M. N. and Studier, F. W. (1977). Analysis of restriction fragments of T7 DNA and determination of molecular weights by electrophoresis in neutral and alkaline gels. *Journal of Molecular Biology* 110, 119–146.

Mead, D. A., Szczesna-Skorupa, E. and Kemper, B. (1986) Single-stranded DNA blue T7 promoter plasmids: a versatile tandem promoter system for cloning and protein engineering. *Protein Engineering* 1, 67–74.

Moukhamedov, R. S., Hu, X., Nazar, R. N. and Robb, J. (1993) Use of PCR-amplified ribosomal intergenic sequences for the diagnosis of *Verticillium tricorpus*. *Phytopathology* (submitted).

Mullis, K., Faloona, F., Scharf, S., Saiki, R., Horn, G. and Erlich, H. (1986) Specific enzymatic amplification of DNA *in vitro*: the polymerase chain reaction. *Cold Spring Harbor Symposia on Quantitative Biology* 51, 263–273.

Nazar, R. N., Hu, X., Schmidt, J., Culham, D. and Robb, J. (1991). Potential use of PCR-amplified ribosomal intergenic sequences in the detection and differentiation of *Verticillium* wilt pathogens. *Physiological and Molecular Plant Pathology* 39, 1–11.

Pegg, G. F. (1974) *Verticillium* diseases. *Review of Plant Pathology* 53, 157–182.

Pegg, G. F. (1981) The impact of some *Verticillium* diseases in agriculture. In: Cirulli, M. (ed.) *Pathobiology of* Verticillium *Species*. Mediterranean Phytopathological Union, Bari, Italy. pp. 82–98.

Pegg, G. F. and Street, P. F. S. (1984) Measurement of *Verticillium albo-atrum* in high and low resistance hop cultivars. *Transactions of the British Mycological Society* 82, 99–106.

Platt, H. W. (1986) Varietal response and crop loss due to *Verticillium* wilt of potato caused by *Verticillium albo-atrum*. *Phytoprotection* 67, 123–127.

Radloff, R., Bauer, W. and Vinograd, J. (1967) A dye-buoyant-density method for the detection and isolation of closed circular duplex DNA: the closed circular DNA in HeLa cells. *Proceedings of the National Academy of Sciences of the United States of America* 57, 1514–1520.

Robb, J., Moukhamedov, R., Hu, X., Platt, H. and Nazar, R. N. (1993) Distinct subgroup niches of *Verticillium albo-atrum, Physiological and Molecular Plant Pathlogy* (submitted).

Rogers, S. O. and Bendich, A. J. (1985). Extraction of DNA from milligram amounts of fresh herbarium and mummified plant tissues. *Plant Molecular Biology* 5, 69–76.

Sanger, F. S., Nicklen, S. and Coulson, A. R. (1977) DNA sequencing with chain-terminating inhibitols. *Proceedings of the National Academy of Sciences of the United States of America* 74, 5463–5467.

Smith, H. C. (1965) The morphology of *Verticillium albo-atrum, V. dahliae* and *Verticillium tricorpus*. *New Zealand Journal of Agricultural Research* 8, 450–478.

Steele, W. S., Okamura, N. and Busch, H. (1965) Effects of thioacetamide on the composition and biosynthesis of nucleolar and nuclear ribonucleic acid in rat liver. *Journal of Biological Chemistry* 240, 1742–1749.

Tuite, J. (1969) *Plant Pathological Methods: Fungi and Bacteria*. Burgess Publishing Co., Minneapolis, Minnesota, USA.

Zhang, H., Scholl, R., Browse, J. and Somerville, C. (1988) Double stranded sequencing as a choice for DNA sequencing. *Nucleic Acids Research* 16, 1220.

13 Detection and Differentiation by PCR of Subspecific Groups Within Two *Verticillium* Species Causing Vascular Wilts in Herbaceous Hosts

J. H. Carder, A. Morton, A. M. Tabrett and D. J. Barbara

Plant Pathology and Weed Science Department, Horticulture Research International, East Malling, West Malling, Kent ME19 6BJ, UK.

Background

Verticillium albo-atrum Reinke and Berthold and *V. dahliae* Klebahn (no teleomorphs known) are important plant pathogens causing vascular wilts in a wide range of annual and perennial dicotyledonous crop and wild plant species around the world. The two species are primarily differentiated by their resting structures (*V. albo-atrum* produces dark resting mycelium while *V. dahliae* produces microsclerotia) but there are other important biological differences such as optimal temperature for growth and pathogenicity. Confusingly, in some previous reports, particularly from the USA, all isolates were referred to as *V. albo-atrum*.

Hops and lucerne are major crops affected by *V. albo-atrum* in northern Europe while potatoes, tomatoes, strawberries and cotton are vulnerable to *V. dahliae* (Tjamos, 1988). Oilseed rape is being increasingly affected by certain host-adapted isolates of *V. dahliae*. Many minor crops are also affected. The fungi are soil-borne but may be disseminated on contaminated seed (e.g. *V. albo-atrum* in lucerne) and control is by disease avoidance, soil fumigation or by breeding resistant or tolerant varieties (e.g. hops, lucerne).

Most isolates of both species are haploid and host non-specific. However, host-adapted haploid isolates of both species do occur, e.g. *V. dahliae* from *Mentha* species (Fordyce and Green, 1960) and *V. albo-atrum* from lucerne. Natural diploids of *V. dahliae* (= var. *longisporum*) are also found and these tend to be pathogens of cruciferous hosts (Jackson and Heale, 1985), particularly oilseed rape.

Molecular studies using RFLPs have shown that both species may be divided into major subspecific groups with little genetic variation within these groups (Carder and

Barbara, 1991; Okoli *et al.*, 1993a,b). Two groups have been found in *V. albo-atrum* (the L group, comprising all isolates tested from lucerne; all other isolates forming the NL group) while four have been found in *V. dahliae* (M, isolates from mint; A and B are distinct but not host-based; D, the diploid isolates). The majority of isolates used in these studies, other than those from mint and some of the diploid isolates, came from the UK or other countries in northern Europe and other subspecific groups may occur elsewhere. At the molecular level the L and NL groups of *V. albo-atrum* and the A and B groups of *V. dahliae* appear to be more closely related to each other than to groups in the other species. The M group seems to be close to the A group of *V. dahliae* while the diploid isolates of this species are at least as different from any of the haploid isolates tested as haploid isolates of *V. dahliae* are from those of *V. albo-atrum* and perhaps should be considered as a separate species. The existence of RFLP groups correlated with the groups of host-adapted isolates was, perhaps, not surprising but the existence of two clear RFLP groups within the non-host-adapted isolates of *V. dahliae* was unexpected. However, these latter groups are probably biologically significant as they are correlated with quantitative differences in pathogenicity to strawberry and form separate vegetative compatibility groups (D. C. Harris and J. R. Yang, personal communication).

Methods used for the control of *V. albo-atrum* and *V. dahliae* may be dependent on the particular subgroup of the two species which is present, e.g. where genes for resistance/tolerance, which may be effective against only some types of isolate, are being incorporated into a crop. Therefore, for both practical and experimental uses, we wanted to develop a rapid and sensitive test capable of detecting and distinguishing between all the subspecific groups identified so far (Carder and Barbara, 1991; Okoli *et al.*, 1993a,b). The primers described in this chapter are capable of identifying some of the RFLP groups and we are now trying to develop others specific for the ones remaining.

A PCR-based assay has been described which uses primers derived from the internal transcribed spacer regions of the ribosomal RNA genes and which is capable of distinguishing *V. albo-atrum* and *V. dahliae* (Nazar *et al.*, 1991; Robb *et al.*, this volume). However, there have been no reports of this assay being assessed with the full range of subspecific groups. The assay described here utilizes primers derived from random genomic probes (Carder and Barbara, 1991; Okoli *et al.*, 1993a,b). This assay has been applied successfully to the detection of isolates of *Verticillium* in several herbaceous hosts.

Development of primers

The subspecific groups were initially defined by RFLPs identified using probes from two partial genomic libraries, one from a non-lucerne isolate of *V. albo-atrum* (Carder and Barbara, 1991) and the second from an A group isolate of *V. dahliae* (Okoli *et al.*, 1993a). Selected clones from these libraries and others produced by cloning repetitive fragments from *Eco*RI digests of total DNA (Morton, Carder and Barbara, unpublished) were sequenced, using standard methods, and a range of primers designed. These primers were assessed for effectiveness and specificity in PCR using highly purified DNAs from a range of *Verticillium* isolates. Primer pairs have been identified which can be used to identify: (i) all isolates of *V. dahliae*; (ii) the NL group of *V. albo-atrum*; and (iii) the A group of *V. dahliae*. Other primers amplify DNA from all isolates (so far tested) of both species. (see Fig. 13.1).

Two rapid methods (Langridge *et al.*, 1991; Edwards *et al.*, 1991) for extracting DNA from fungal and plant tissues that have been used successfully on tomato, lucerne and *Arabidopsis* were not routinely successful on strawberry or hop tissue and other extraction procedures, suitable for use with these and other 'difficult' hosts, are now being investigated.

PROTOCOLS

DNA extraction

Both the procedures given are applicable to infected plant tissue and to mycelium taken from plate or liquid culture.

Method 1

Squash blot (slightly adapted from Langridge *et al.*, 1991). [Only Hybond-N (Amersham plc) is reported to be a suitable membrane.]

1. Prewet membrane in 1 M NaOH and lay on to 3 MM paper, similarly prewetted, on a glass sheet.
2. Crush sample on to membrane using disposable plastic stirring rod (we routinely use petiole or stem pieces *ca.* 2 mm^3).
3. Rinse membrane in 1.5 M NaCl, 0.5 M Tris, 1 mM EDTA, pH 7.
4. Rinse membrane three times in 10 mM Tris-HCl, 1 mM EDTA, pH 7.5.
5. Rinse membrane in water. Membranes can now be air dried and stored or used immediately.
6. Cut out areas of membrane with sample and heat with 60 μl water in PCR tube for 5 min at 94°C. Use 7 μl for PCR.

Method 2

SDS extraction. This method has the advantage of providing more DNA than the preceding technique but involves rather more handling of samples. Yield of DNA is *ca.* 50–100 ng per sample.

1. Grind 2 mm^3 stem or petiole sample in microcentrifuge tube (we use Kontes disposable pellet pestles obtained through Hoefer Scientific Instruments) with 400 μl extraction solution (0.2 M Tris-HCl pH 7.5, 0.25 M NaCl, 25 mM EDTA, 5 g l^{-1} SDS).
2. Vortex 5 s. Samples can be left at room temperature until all are extracted.
3. Centrifuge, 13,000 rpm for 1 min.
4. Mix 300 μl supernatant added to 300 μl isopropanol in a fresh microcentrifuge tube; stand 2 min.
5. Centrifuge 13,000 rpm for 5 min and discard supernatant.
6. Dry pellet and resuspend in 100 μl 10 mM Tris-HCl, 1 mM EDTA, pH 8.0.
7. Estimate DNA concentration by agarose gel electrophoresis. We routinely use 1–2 μl for each amplification.

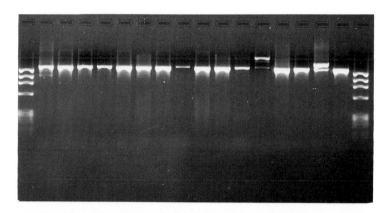

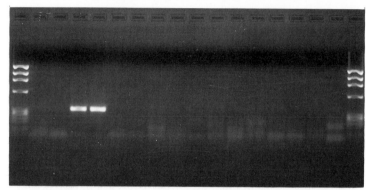

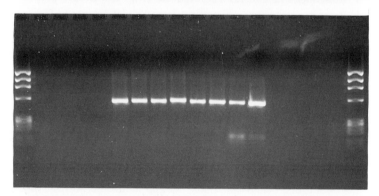

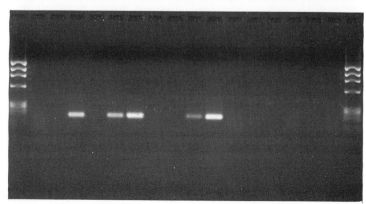

Primers

The most useful primers and the range of isolate types they amplify are listed below:

53/54   5'-CAT GGA TAA CCG TGG TAA TT-3'
         5'-CCA TTC AAT CGG TAG TAG CG-3'
         Product *ca.* 1500 bp except some diploid isolates of *V. dahliae ca.* 2500 bp.
         Amplifies DNA from isolates of all *Verticillium* species tested, including some
         other than *V. dahliae* and *V. albo-atrum*, but distinguishes some diploid isolates
         by virtue of a larger product size. Not useful for *in planta* testing as they were
         derived from the ribosomal RNA gene region and amplify plant DNA.

2/3   5'-ATG GAC CGA ACA GCT AGG TA-3'
         5'-TCT CAG ATA TAT GCT GCT GC-3'
         Product 300 bp. Amplifies NL group of *V. albo-atrum* but not L group or *V.
         dahliae* (except some mint isolates rather poorly).

19/22   5'-CGG TGA CAT AAT ACT GAG AG-3'
         5'-GAC GAT GCG GAT TGA ACG AA-3'
         Product *ca.* 580 bp. Amplifies all isolates of *V. dahliae*, including diploid ones,
         but not *V. albo-atrum*.

42/70   5'-GTT TCT TAG CTT GCA ACA T-3'
         5'-ACG AGA GTG GAA TAA AGC GA-3'
         Product *ca.* 190 bp. Amplifies RFLP group A isolates and mint and some diploid
         isolates of *V. dahliae* but not group B isolates and only some *V. albo-atrum* NL
         group isolates.

PCR

For amplification, we use standard conditions with 1.5 mM MgCl$_2$. Reactions are
normally 25 μl containing 10 pmol each primer, 0.8 μm dNTPs (Pharmacia) and 0.5 U
Taq polymerase (usually BRL) and overlaid with *ca.* 30 μl mineral oil. Routinely, we run 35
cycles with an initial annealing for 2 min, then 93°C/1 min annealing tempera-
ture/1 min-72°C/1 or 2 min and a final 72°C for 5–10 min. Annealing temperatures used
are 4°C below the lowest T_m of the primer pair (calculated according to 4(G + C) +
2(A + T)) although we have used 48°C successfully with most of these primers. After
amplification 10 μl are run out in a 15 g l^{-1} agarose gel (Bio-Rad, DNA grade), 15 cm long
(usually with three combs allowing *ca.* 60 samples per gel) in TBE with 0.5 μg ml^{-1}

Fig 13.1.   Amplification products from four primer pairs with a range of *Verticillium*
isolates. Primers pairs are: top row 53/54, second row 2/3, third row 19/22, bottom row 42/70.
Samples are: (lane 1) MW markers (φX174 RF DNA/*Hae*III fragments); (lane 2) *Verticillium albo-
atrum* 220 and (lane 3) 235 (both RFLP group L); (lane 4) *V. albo-atrum* 1974 and (lane 5) 1844
(both NL group); (lane 6) *V. dahliae* 2341 and (lane 7) 122 (both RFLP group A); (lane 8) *V. dahliae*
111 and (lane 9) Ω (both RFLP group B); (lane 10) *V. dahliae* A1 and (lane 11) IMI (both RFLP
group M); (lane 12) *V. dahliae* 84120 and (lane 13) 162 (both diploid isolates); (lane 14) *V.
nigrescens* 1879; (lane 15) *V. tricorpus* 1988; (lane 16) *V. nubilum* 1881; (lane 17) *V. lecanii* 1; (lane
18) MW markers. Primers 53/54 amplify DNA from all isolates of *Verticillum* and from plants (data not
shown); 2/3 *V. albo-atrum* NL group; 19/22 all isolates of *V. dahliae*; 42/70 amplify *V. dahliae*
subgroup A$_1$ and subgroup NL$_1$ of *V. albo-atrum* but not *V. dahliae* group B or subgroup A$_2$ of this
species (only one isolate has been identified in this subgroup) or NL$_2$ of *V. albo-atrum*. (For details of
the subgroups see Okoli *et al.*, 1993a.)

ethidium bromide for *ca.* 1 h at 150 V. These conditions give good resolution for fragments above *ca.* 500 bp. A gel composed of $30 \, g \, l^{-1}$ Nusieve GTG agarose (FMC) with $10 \, g \, l^{-1}$ Bio-Rad agarose (in TBE/ethidium bromide), which gives better resolution of smaller fragments, may also be used.

Sensitivity

The limit of detection of purified fungal DNA is $\leqslant 1$ pg ($\leqslant 45$ genome copies) or *ca.* 200 pg of total DNA from infected plants ($\simeq 80 \, \mu g$ of tissue).

LIMITATIONS

Due to the high sensitivity of PCR techniques they are especially susceptible to problems of contamination. We have found this a particular problem when handling recombinant plasmids for RFLP work and PCR primers derived from them in the same general laboratory area. This has been overcome by strict separation of working areas and the use of laminar flow cabinets for setting up PCR reaction mixtures. We have also encountered inhibition of amplification when using water sterilized by DEPC treatment or, more surprisingly, by autoclaving in blue-plastic capped glass bottles.

Few of the primers described are entirely specific for individual RFLP groups but combinations of pairs of primers will differentiate several groups. The lack of primers specific for some groups may not be a problem in practice as several of them appear to be host-specific.

REFERENCES

Carder, J. H. and Barbara, D. J. (1991) Molecular variation and restriction fragment length polymorphisms (RFLPs) within and between six species of *Verticillium*. *Mycological Research* 95, 935–942.

Edwards, K., Johnstone, C. and Thompson C. (1991) A simple and rapid method for the preparation of plant genomic DNA for PCR analysis. *Nucleic Acids Research* 19, 1349.

Fordyce, C. and Green, R. J. (1960) Studies on the host specificity of *Verticillium albo-atrum* var. *menthae*. *Phytopathology* 50, 635.

Jackson, C. W. and Heale, J. B. (1985) Relationship between DNA content and spore volume in sixteen isolates of *Verticillium lecanii* and two new diploids of *V. dahliae* (= *V. dahliae* var. *longisporum*). *Journal of General Microbiology* 131, 3229–3236.

Langridge, U., Schwall, M. and Langridge, P. (1991) Squashes of plant tissue as substrate for PCR. *Nucleic Acids Research* 19, 6954.

Nazar, R. N., Hu, J., Schmidt, J., Culham, D. and Robb, J. (1991) Potential use of PCR-amplified ribosomal intergenic sequences in the detection and differentiation of verticillium wilt pathogens. *Physiological and Molecular Plant Pathology* 39, 1–11.

Okoli, C. A. N., Carder, J. H. and Barbara, D. J. (1993a) Molecular variation and sub-specific groupings within *Verticillium dahliae*. *Mycological Research* 97, 233–239.

Okoli, C. A. N., Carder, J. H. and Barbara, D. J. (1993b) Restriction fragment length poly-morphisms (RFLPs) and the relationships of some host-adapted isolates of *Verticillium dahliae*. *Plant Pathology* (in press).

Tjamos, E. C. (1988) *Verticillium dahliae* Kleb. and *Verticillium albo-atrum* Reinke and Berthold. In: Smith, I. M., Dunez, J., Phillips, D. H., Lelliott, R. A. and Archer S. A. (eds) *European Handbook of Plant Diseases*. Blackwell Scientific, Oxford, pp. 299–302.

14 A Double (Monoclonal) Antibody Sandwich ELISA for the Detection of *Verticillium* Species in Roses

M. M. van de Koppel and A. Schots

Laboratory for Monoclonal Antibodies, PO Box 9060, 6700 GW Wageningen, The Netherlands.

Background

Verticillium dahliae and *Verticillium albo-atrum* cause vascular wilt disease in a wide range of economically important crops, ornamentals, weeds and native plants all over the world. Important host plants are potato, strawberry, cotton, eggplant, tomato, pea, maple, cherry, elm and rose (Visser and Kotzé, 1975; Isaac, 1967; Isaac and Harrison, 1968; Woolliams, 1966).

Verticillium species cause wilting by blocking the xylem vessels of the plant. The fungus survives in the soil as microsclerotia, initially embedded in fragments of plant tissue. These resting structures can survive for years. Root exudates stimulate them to germinate after which they penetrate directly into the vascular system of the plant (Lacy and Horner, 1966; Woolliams, 1966; Isaac, 1967; Evans and Gleeson, 1973). At an early stage, when small amounts of mycelium are present, the plant does not show symptoms although at a later stage, when symptoms appear, the amount of mycelium is greater and the fungus begins to form conidia (Isaac and Harrison, 1968).

Characteristic symptoms of *Verticillium* wilt are wilting at the tips of young leaves and a yellowing of lower leaves. After a few days permanent wilting may occur. The leaves turn yellow and brown and finally drop (Horst, 1979, 1983). Symptoms appear during periods of stress, such as drought in midsummer. Seasonal variation is less evident in glasshouses, where the pathogen may be active during the whole period of planting. Because the symptoms show up at a late stage, it is hard to distinguish them from the normal symptoms of ageing plants (Isaac and Harrison, 1968). A potential complication is that infected plants may tolerate the disease when growing conditions are optimal. Moreover, after irrigation or a fall in temperature the plants can recover again. In these cases, *Verticillium* can be isolated from infected tissue even though symptoms are lacking. With a return to drier conditions wilting can occur again, leading to death of the plant. *Verticillium* wilt is a major problem for field and glasshouse grown roses. The disease is very difficult to control after it has appeared. Infected plants must be removed and the soil sterilized. Infected but symptomless plants delivered by a supplier often only show symptoms after being planted by the grower (Horst, 1979, 1983; Isaac and Harrison, 1968; Woolliams, 1966). Therefore it is very important to use effective detection methods before planting. Conventional methods, such as growing out of the fungi on agar demand a lot of time and expertise and bioassays for the detection of *Verticillium* are not always reliable (Woolliams, 1966). This study aims to investigate whether a serological test with monoclonal antibodies can be used for rapid detection of *V. dahliae* (and also *V. albo-atrum*) at an early stage and the differentiation between this and less pathogenic fungi in roses. An immunoassay for *V. dahliae* in roses might also be suitable for the detection of this fungus in other plants.

Background to immunoassays for *V. dahliae*

Several polyclonal antibody-based immunoassays for *Verticillium* have been developed, mostly for the detection of *V. dahliae*. These antibodies were used to compare isolates of *V. albo-atrum* and *V. nigrescens* (Wyllie and DeVay, 1970) and to determine the antigenic similarities between different *Verticillium* species (Fitzell *et al.*, 1980). Charudattan and DeVay (1972) investigated common antigens among varieties of cotton and isolates of *Fusarium oxysporum* f.sp. *vasinfectum, F. solani* f.sp. *phaseoli, V. albo-atrum* and *V. nigrescens*. All these authors developed antibodies against different isolates of fungi and tested the reactions with other isolates and with plant material to determine the common antigens. Our aim was to find antibodies that could be used to detect *V. dahliae* and *V. albo-atrum* and that do not react with common antigens from other (non-pathogenic) fungi or from plant material.

For the development of these antibodies mice were immunized with surface washings and culture supernatant. Fusions were carried out (Schots *et al.*, 1992) which resulted in nine hybridoma cell-lines producing antibodies that recognized *V. dahliae* in roses, and reacted with surface washings and culture supernatant of *V. dahliae*. They did not react with non-inoculated plants. One of these clones, 7E12 which is of the IgG2a subclass, was used for further ELISA tests. The antibodies could be diluted further and the OD readings were higher when compared to the other clones. With this clone different ways to carry out the ELISA protocols were investigated. A DAS-ELISA turned out to be the best method. Plates were coated with monoclonal antibodies from 7E12 and alkaline phosphatase conjugates (Tijssen, 1985) of the MAb-7E12 were used as second antibodies.

Production of monoclonal antibodies to *V. dahliae*

Antigens for immunization of the mice and screening of hybridoma supernatants were obtained from surface washings and culture supernatants from *Verticillium dahliae* grown in the dark on potato dextrose agar slants and in Czapeck's Dox (CD) liquid cultures.

Surface washings were obtained by washing the slants with 2 ml Milli-Q water and then gently stroking the surface with a Pasteur pipette. The suspension was centrifuged and the supernatant was concentrated 10 times in an Amicon Ultrafiltration Cell using a filter with a cut off of 10 kD.

Culture supernatant was obtained from culture fluid after removal of the mycelium by filtering through cheese cloth followed by removal of the conidia by centrifugation for 10 min at 1500g. After filtering, the high molecular glycoproteins were removed from the supernatant using a 0.65 μm filter. Then the supernatant was concentrated with a Filtron ultrafilter with a 5 kD cut off to 100 μg ml^{-1}.

PROTOCOL

Materials

1. Flatbottom polystyrene microtitre plates (Costar, Labstar 6891/S)2.
2. Antigen extraction buffer: PBS+0.1% Tween 20 + 4% PVP (polyvinylpyrrolidone).
3. Coating buffer: 50 mM Na$_2$CO$_3$/NaHCO$_3$ pH 9.6.
4. Washing buffer: PBST.
5. Blocking buffer: PBST + 0.5% BSA.
6. Conjugate buffer: PBST + 0.1% BSA.
7. Substrate solution: 7.5 mg *p*-nitrophenylphosphate in 10 ml substrate buffer (2 M diethanolamine, pH 9.8).

Method

1. Grind the plant material in a blender with extraction buffer 1:2, centrifuge the extracts for 5 min at 1500g and collect the supernatant. Use direct or freeze at −20°C.
2. Coat the wells of the plate with the monoclonal antibodies, diluted 1:3000 in coating buffer, 100 μl per well.
3. Incubate 2 h at 37°C or overnight at 4°C.
4. Wash twice with washing buffer.
5. Incubate the plate with blocking buffer, 200 μl per well for 30 min at 37°C.
6. Wash the plate twice with washing buffer.
7. Incubate with plant extracts 100 μl per well, 2 h at 4°C.
8. Wash the plate twice with washing buffer.
9. Incubate with the AP-conjugated monoclonal antibodies 100 μl per well 1:5000 in conjugate buffer, 1 hour at 4°C.
10. Wash the plate four times with washing buffer.
11. Add 100 μl substrate solution to each well and read the absorbance at 405 nm in an ELISA reader.

RESULTS

The monoclonal DAS-ELISA was used to detect *V. dahliae* in infected plants. Roses were inoculated with conidia from cultures of *V. dahliae* grown on potato. Stem parts of the plants taken at different heights, 0–10, 10–25, 25–40 and 40–60 cm, were tested to see how the fungus had spread through the plant. Six of the ten inoculated roses reacted (2, 3, 4, 5, 8 and 9), most of them only in the lower parts of the plant and not in the leaves. No reaction was found with the non-inoculated plants (Figs 14.1 and 14.2).

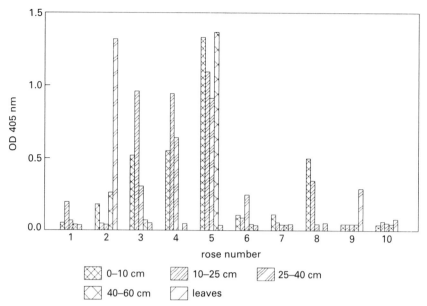

Fig 14.1. DAS-ELISA with ten roses inoculated with *V. dahliae*. Coating: monoclonal antibodies (7E12), 1 : 3000 diluted. Antigens: extracts from plants at different heights (0–10 cm, 10–25 cm, 25–40 cm, 40–60 cm and the leaves). Antibodies: 7E12-AP, 1 : 5000 diluted. Incubation with substrate 3 h.

The ELISA was also tested with chrysanthemums which were inoculated with *V. dahliae* or *V. albo-atrum*. The results of the ELISA were correlated to the symptoms of the plants and growth of the fungus on agar. In some samples the ELISA was negative where the plant showed symptoms or where *Verticillium* was found on the agar. However, the opposite never occurred: when symptoms were not observed or where *Verticillium* did not grow from stem segments in agar platings the ELISA never gave a positive reaction. The failure to detect every infection is probably due to the fact that we used extracts from the whole plant instead of from only the lowest part of the stem. This could have diluted the antigen. When extracts are made from whole plants the antigen may be diluted below the detection levels. Another cause of negative ELISA results could be the use of too large a volume of extraction buffer making the antigen too dilute to detect.

Finally, the ELISA was tested with roses from growers with *Verticillium* problems in their crops. Plants with clear symptoms did not react by ELISA. All the plants from which *V. dahliae* was isolated were positive in the ELISA. Where the ELISA results were negative *V. dahliae* never grew on the agar. However, we did not always find *V. dahliae* on the agar plates with every sample that was positive in the ELISA. In these cases there were many different other fungi on the plate that could have masked the presence of *V. dahliae*.

LIMITATIONS

With this assay we were able to detect *V. dahliae* in roses inoculated with the pathogen and in naturally infected roses, and *V. dahliae* and *V. albo-atrum* in inoculated

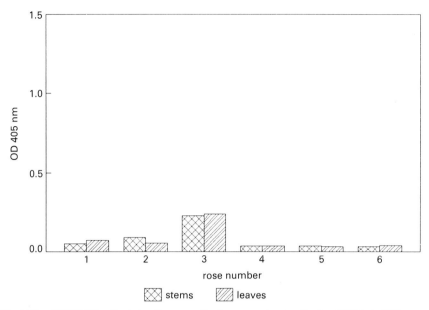

Fig 14.2. DAS-ELISA with six healthy roses. Coating: monoclonal antibodies (7E12), 1 : 3000 diluted. Antigens: extracts from the stems and the leaves. Antibodies: 7E12-AP, 1 : 5000 diluted. Incubation with substrate 3 h.

chrysanthemums. We did not get cross-reactions with healthy plant material, but more research needs to be done on cross-reactions with other fungi in roses. The usefulness of the assay also needs to be tested with more isolates of *V. dahliae* (and other *Verticillium* species) and more cultivars and species of roses and other plant species infected with *V. dahliae*.

REFERENCES

Charudattan, R. and DeVay, J. E. (1972) Common antigens among varieties of *Gossypium hirsutum* and isolates of *Fusarium* and *Verticillium* species. *Phytopathology* 62, 230–234.

Evans, G. and Gleeson, A. C. (1973) Observations on the origin and nature of *Verticillium dahliae* colonizing plant roots. *Australian Journal of Biological Science* 26, 151–161.

Fitzell, R., Fahy, P. C. and Evans, G. (1980) Serological studies on some Australian isolates of *Verticillium* spp. *Australian Journal of Biological Science* 33, 115–124.

Horst, R. K. (1979) Verticillium. In: *Westcott's Plant Disease Handbook* 4th edn. Revised by R. K. Horst, Van Nostrand Reinholt Co., New York, pp. 439–443.

Horst, R. K. (1983) *Compendium of Rose Diseases*. American Phytopathological Society, St Paul, Minnosota, USA pp. 12–13.

Isaac, I. (1967) Speciation in *Verticillium. Annual Review of Phytopathology* 5, 201–221.

Isaac, I. and Harrison, J. A. C. (1968) The symptoms and causal agents of early-dying disease *Verticillium* wilt) of potatoes. *Annals of Applied Biology* 61, 231–244.

Lacy, M. L. and Horner, C. E. (1966) Behaviour of *Verticillium dahliae* in the rhizosphere and on roots of plants susceptible, resistant, and immune to wilt. *Phytopathology* 56, 427–430.

Schots, A., Pomp, R. and Van Muiswinkel, W. B. (1992) Production of monoclonal antibodies. In: Stolen, J. S., Fletcher, T. C., Anderson, D. P., Kaattari, S. L. and Rowley, A. F. (eds) *Techniques in Fish Immunology*. SOS Publications, Fair Haven, NJ, pp. 1–18.

Tijssen, P. (1985) Preparation of enzyme–antibody or other enzyme–macromolecule conjugates. In: Burdon, R. H. and van Knippenberg, P. H. (eds) *Laboratory Techniques in Biochemistry and Molecular Biology*, vol. 15. *Practice and Theory of Enzyme Immunoassays*. Elsevier, Amsterdam, New York, Oxford, pp. 221–278.

Visser, S. and Kotzé, J. M. (1975) Reactions of various crops to inoculation with *Verticillium dahliae* Kleb. isolated from wilted strawberry plants. *Phytophylactica* 7, 25–30.

Woolliams, G. E. (1966) Host range and symptomatology of *Verticillium dahliae* in economic, weed, and native plants in interior British Columbia. *Canadian Journal of Plant Science* 46, 661–669.

Wyllie, T. D. and DeVay, J. E. (1970) Immunological comparison of isolates of *Verticillium albo-atrum* and *V. nigrescens* pathogenic to cotton. *Phytopathology* 60, 1682–1686.

15 Detection of *Heterobasidion annosum* Using Monoclonal Antibodies

D. N. GALBRAITH AND J. W. PALFREYMAN

Scottish Institute for Wood Technology, Dundee Institute of Technology, Bell Street, Dundee DD1 1HG, UK.

BACKGROUND

Heterobasidion annosum (Fr.)Bref. is one of the most common basidiomycete organisms responsible for the decay of conifer species in temperate zones (Hallaksela, 1984). The organism causes a heart and root rot which is the most economically important disease of coniferous trees (Cartwright and Findlay, 1958; Redfern, 1982). Infection of susceptible trees can be by spore infection of stump material followed by transfer of mycelium through root contacts or, rarely, by direct infection by spores through bark wounds (Rishbeth, 1951; Hodges, 1969).

The effect of the fungus on trees is related to a number of factors including the species of tree infected and the strain of the *H. annosum*. For example *H. annosum* strain S causes root rot and heartrot in Norway spruce and can kill Scots pine saplings but has not been recovered from mature pines (Korhonen, 1978). By contrast, strain P has been isolated from pines and also a range of other woody plants not attacked by the S strain. The third European strain (F) is apparently limited in both its geographical and host range to the common silver fir of the Apennine mountains and Italian Alps (Capretti *et al.*, 1990). The first visible sign of infection caused by *H. annosum* is discoloration of irregularly shaped zones in the wood at the base of the tree. Eventually such zones develop into a stringy, dry decay. This decay can spread within the heartwood of the tree resulting in increased susceptibility to windthrow (Greig and Redfern, 1974) and loss of commercial value of timber harvested from the tree (Hodges, 1969).

Currently, detection of *H. annosum* infection is achieved by knowledge of plantation history, the development of fruit bodies in the forest floor and the isolation of the organism from cores. In addition, the detection of conidiospores on timber discs cut from freshly felled trees is indicative of infection by the organism. The development of simple assays for *H. annosum* would be useful for a variety of reasons, for example: (i) the degree of infection of specific forests and the spread of disease through such forests could be easily monitored; (ii) the mapping of organism colonization of specific trees (currently achieved by microscopic inspection of incubated wood discs) could be facilitated; (iii) disputes which occur at times between foresters and sawmills over the nature of staining found in harvested timber could be resolved more easily. In addition the development of simple test systems for the identification of strains of *H. annosum* would allow the spread of the strains and their host affinities to be more easily evaluated.

Antibody systems

To date, immunological methods have not been widely applied to the detection of *H. annosum* though the development of polyclonal antisera against the organism was reported by Avramenko (1989). While such antisera can be useful in identifying organisms by Western blotting (Vigrow *et al.*, 1991), their application to simple assay systems such as the ELISA is limited if cross-reactivity is extensive. In this chapter the development of a monoclonal-antibody-based assay for *H. annosum* is described as is an attempt to produce a strain-specific monoclonal antibody using immunomodulators.

PROTOCOL

Monoclonal antibody-based assay for the detection of *H. annosum*

Monoclonal antibodies were produced by standard techniques based on those reported by Dewey *et al.* (1990) using surface washings (0.5 ml of a 5 ml PBS extract from an 85 mm plate) as the immunogen. The first two immunizations (intraperitoneal) were given at 2 week intervals, after testing sera by ELISA suitable animals were immunized intravenously 4 days prior to the fusion experiment. Fungi used in the study were obtained from a variety of sources including the Forestry Authority (UK), Dr K. Korhonen (Finnish Forest Research Institute, Vantaa, Finland), Dr J. Karlsson (Swedish University of Agricultural Sciences, Uppsala), Building Research Establishment (UK) and the International Mycological Institute (Kew, UK). Production of 'S'-strain-specific monoclonal antibodies was by modification of the method of Hamilton *et al.* (1990). Mice were immunized intraperitoneally with an exoantigen preparation from *H. annosum* strain FC 51.8 (a P isolate, amounts as described above) followed by cyclophosphamide (40 mg kg^{-1} body weight) at 15 min, 24 and 48 h. This procedure was repeated four times at 15 day intervals. Twenty-one days after the last immunization the mice were given an immunization with *H. annosum* strain FIN.9 (an S isolate). After a final boost with the S strain isolate mice were bled and those producing high antisera titres in the appropriate ELISA were selected as potential spleen donors for hybridoma production.

For assay of infected wood samples or of tree cores, material was milled to a fine sawdust (<1 mm particles) with a Culatti DFH48 hammer mill, incubated overnight with PBS at 4°C with agitation (1:2 w/v), centrifuged at 12,000 g for 10 min and the supernatant used as the antigen source in ELISA or dot blot immunoassay.

A standard ELISA procedure was used for assay of both hybridoma supernatants and infected wood extracts. Extracts were incubated overnight in microtitre plate wells, blocked with PBS/Tween/new born calf serum (NCS) (0.5%/10%) and incubated with the appropriate monoclonal antibody supernatant. Antibody binding was detected by using horseradish peroxidase linked antimouse immunoglobulin (Scottish Antibody Production Unit (SAPU)) diluted 1:500 in PBS/Tween (0.05%) followed by the chromagen 3,3′,5,5′-tetramethylbenzidine.

For the dot blot assay 100 μl aliquots of extracts were transferred to the appropriate wells of an AttO AE-6190 microfiltration apparatus loaded with a sheet of PVDF Immobilon pre-wetted with methanol and PBS/Tween (0.05%). Gentle suction was applied to draw the liquid through the apparatus and to bind the antigen to the membrane. Antigen dots were washed (×6) with 400 μl PBS/Tween (0.05%), blocked with PBS/Tween/NCS (0.5%/10%) for 60 min and incubated with the appropriate monoclonal antibody supernatant overnight at 4°C. After six washes with PBS/Tween (0.05%), horseradish peroxidase linked antimouse immunoglobulin (SAPU) diluted 1:500 in PBS/Tween/NCS (0.05%/5%) was added and incubated for 60 min at room temperature. After a further six washes dots were developed using 3,3′-diaminobenzidine/nickel chloride (Harlow and Lane, 1988). Included in all assays were antigen preparations containing 100, 50, 25, 12.5 and 0 μg of *H. annosum* FC 51.8 freeze-dried mycelium as standards. The test dots were compared visually with the standards and scored '4' if comparable to the 100 μg dot, '3' if comparable with the 50 μg dot and so on. Dots of lesser intensity than the 12.5 μg standard were scored '0'.

Antibody specificity

The specificities of monoclonal antibodies produced by the standard immunization schedule (MAb 2/6/D4) and the cyclophosphamide modified schedule (MAb 1/10/D5) are shown in Fig. 15.1(a) and (b) respectively. No organisms tested, including those shown in Fig. 15.1(a) plus a further unidentified 42 isolates from pine trees, reacted with monoclonal antibody 2/6/D4 other than *H. annosum*. All 37 strains of *H. annosum* tested (including representatives of the P, S and F strains) reacted with the monoclonal antibody (data not shown). Monoclonal antibody 1/10/D5, intended to be specific to the S strain, did not react with any non-*H. annosum* organisms (data not shown) but reacted with all S isolates tested and three of six P isolates (Fig. 15.1(b)). Both antibodies were IgG subclass 1 with kappa light chains.

FIELD SAMPLE RESULTS

ELISA and dot blot assay results of extracts from cores taken from a forest severely infected with *H. annosum* are shown in Fig. 15.2(a) and (b) respectively. As well as immunoassays the presence of appropriate conidia on isolates made from a further set of cores was determined. Conidia, ELISA and dot blot results indicated a significant, though not perfect, correlation between the three identification systems. Analysis of samples from a second forest, thought to be uninfected, produced a much lower number of ELISA and dot blot positive scores though the assay revealed that the forest did indeed contain some infected trees.

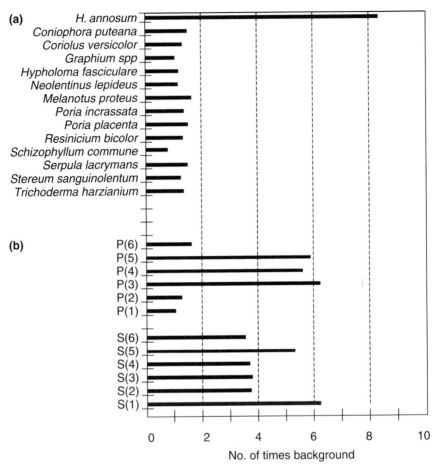

Fig 15.1. (a) Graph indicating the cross-reactivity of MAb 2/6/D4 with other fungi. (b) Cross-reactivity of antibody MAb 1/10/D5 with a range of P and S isolates of *H. annosum.*

LIMITATIONS/OPPORTUNITIES

At present the assays described in this chapter cannot be used on site and are relatively slow, however they do demonstrate the applicability of immunoassays to the detection of *H. annosum.* In general the dot blot assay gave more positive results than the ELISA which gave more positives than detection of conidia. The explanations for these discrepancies are not yet apparent though: (i) no organisms other than *H. annosum* reacted with monoclonal antibody 2/6/D4 used in the assay; (ii) the antibody did not interact with uninfected wood; and (iii) a survey of two forests, one widely infected with *H. annosum* the other reportedly uninfected, gave more positive results in the former forest than in the latter.

It may well be that the immunoassays are more sensitive than the isolation analysis or that exoantigens are more widespread in infected trees than in areas of actively growing

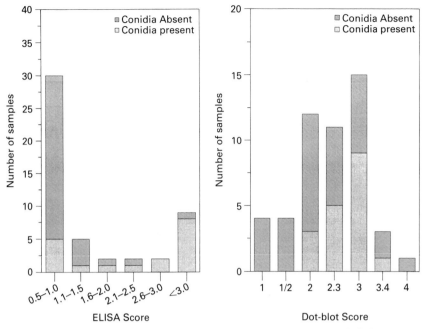

Fig 15.2. (a) Summary of results for Forest 1 comparing ELISA score with mycological examination. (b) Summary of results for Forest 1 comparing dot blot score with mycological examination.

fungus. Further evidence for the specificity of the immunoassays comes from the finding that in all samples assayed where *H. annosum* or an exoantigen was present, positive results were obtained. However, further analysis will be required before the usefulness of the reagents developed in this project can be fully ascertained.

Despite repeated use of the immunomodulator cyclophosphamide, a strain-specific monoclonal antibody was not produced in this study. However, in contrast to the highest specificity antibody produced by conventional techniques (2/6/D4) some strain specificity was apparent and further use of immunomodulators, perhaps combined with partially purified antigens, might allow the production of fully strain-specific reagents. Alternatively, it may be that the members of the P group tested in this analysis were not homogeneous and this possibility will be tested by other means.

ACKNOWLEDGEMENTS

Thanks are due to the Scottish Forestry Trust and the Forestry Authority for their financial support for this work and to Dr Derek Redfern (Northern Research Station of the Forestry Authority) for his guidance in the development of the project and in the practical aspects of the work associated with *H. annosum* isolation, culture and morphological identification.

REFERENCES

Avramenko, R. S. (1989) A serological study of strains of *Heterobasidion annosum* (Fr.) Bref. 1. *Heterobasidion annosum* from the common pine. *Mikologiya I Fitopatologiya* 23, 225–233.

Capretti, P., Korhonen, K., Mugnai, L. and Romognoli, C. (1990) An intersterility group of *Heterobasidion annosum* specialised to *Abies alba*. *European Journal of Forest Pathology* 20, 231–240.

Cartwright, K. St. G. and Findlay, W. P. K. (1958) *Decay of Timber and its Prevention*. Department of Scientific and Industrial Research, HMSO, London.

Dewey, F. M., MacDonald, M. M., Philips, S. I. and Priestley, R. A. (1990) Development of monoclonal-antibody-ELISA and -DIP-STICK immunoassays for *Penicillium islandicum* in rice grains. *Journal of General Microbiology* 136, 753–760.

Greig, B. J. W. and Redfern, D. B. (1974) *Fomes annosus*. Forestry Commission Leaflet No. 5, 1–9.

Hallaksela, A. (1984) Causal agents of butt-rot in Norway spruce in southern Finland. *Silva Fennica* 18, 237–243.

Hamilton, A. J., Bartholomew, M. A., Fenelon, L. E., Figuerora, J. and Hay, R. J. (1990) A murine monoclonal antibody exhibiting high specificity for *Histoplasma capsulatum* var. *capsulatum*. *Journal of General Microbiology* 136, 331–335.

Harlow, E. and Lane, D. (1988) Immunoblotting. In: *Antibodies: A Laboratory Manual*. Cold Spring Harbor, New York, pp. 471–510.

Hodges, C. S. (1969) Modes of infection and spread of *Fomes annosus*. *Annual Review of Phytopathology* 7, 247–266.

Korhonen, K. (1978) Intersterility groups of *Heterobasidion annosum*. *Communications of the Institute of Forest Fenn.* 94, 1–25.

Redfern, D. B. (1982) Infection of *Picea sitchensis* and *Pinus contorta* stumps by basidiospores of *Heterobasidion annosum*. *European Journal of Forest Pathology* 12, 11–25.

Rishbeth, J. (1951) Observations on the biology of *Fomes annosus* with particular reference to East Anglian pine plantations. II. Spore production, stump infection and saprophytic activity in stumps. *Annals of Botany* 25, 221–246.

Vigrow, A., King, B. and Palfreyman, J. W. (1991) Studies of *Serpula lacrymans* mycelial antigens by Western blotting techniques. *Mycological Research* 95, 1423–1428.

16 Identification of *Heterobasidion annosum* (Fr.) Bref., a Root and Butt Rot Disease of Trees, by PCR Fingerprinting

R. KARJALAINEN AND K. KAMMIOVIRTA

Department of Plant Biology, Plant Pathology Section, University of Helsinki, 00014 Helsinki, Finland.

BACKGROUND

Heterobasidion annosum (Fr.) Bref. is a serious root and butt rot pathogen of conifers throughout the northern temperate regions of the world (Hodges, 1969; Chase and Ullrich, 1988). *H. annosum* is divided into different intersterility groups which have different host preferences (Korhonen, 1978). In Europe, the S group infects mainly *Picea abies*, the F group *Abies alba*, while the P group prefers *Pinus sylvestris* but can also attack many other other trees (Korhonen, 1978; Capretti *et al.*, 1990). S and P group isolates are widely distributed in Europe but F group occurs mainly in northern Italy. Similar intersterility (IS) groups have been found in North America but they are only partially compatible with their corresponding European groups (Chase and Ullrich, 1988; Harrington *et al.*, 1989). Intersterility between groups is controlled by at least five IS genes (Chase and Ullrich, 1988).

Identification of intersterility groups of *H. annosum* is of considerable importance because of their differing host preference. Identification of IS groups can be performed by pairing tests; the pathogen is heterothallic and mated secondary hyphae can form clamp connections (Korhonen, 1978). However, this is a time-consuming and laborious task. Recently, isoenzyme analysis (Otrosina *et al.*, 1992) and pectinase profiles (Karlsson and Stenlid, 1991) have been used to differentiate IS groups of *H. annosum*. We show here that the amplification of *Heterobasidion* DNA by the RAPD-PCR method is a reliable and convenient way to identify different isolates of this pathogen.

PROTOCOL

Materials

Fungal culturing

1. 1.5% malt agar (20 g malt powder, 15 g agar in 1 l H$_2$O).
2. Cellophane film 400P (Ahlström Eurapak).

DNA extraction

1. Lysis buffer: 50 mM Tris-HCl pH 7.2; 50 mM EDTA pH 8.0; 3% SDS; 1% 2-mercaptoethanol.
2. Chloroform.
3. Phenol (equilibrated to pH 8.0).
4. 3 M sodium acetate pH 8.0.
5. Isopropanol.
6. Ethanol 70%.
7. TE: 10 mM Tris-HCl pH 8.0, 0.1 mM EDTA.
8. RNase A 10 mg ml^{-1}.

PCR

1. 10 × buffer: 0.1 M Tris-HCl pH 8.3; 1 mg ml^{-1} BSA; 0.5 M KCl; 10–60 mM MgCl$_2$ (depends on the primer optimum).
2. Polymerase (Dyna-Zyme from Finnzymes or *Taq* from Boehringer).
3. dNTPs (dATP, dTTP, dGTP, dCTP).
4. Sterile paraffin oil.

RAPD primers

The nucleotide sequences of primers were synthesized using the Applied Biosystems DNA synthesizer (model 381A) at the Biotechnology Institute of the University of Helsinki. The sequences and their optimal MgCl$_2$ concentrations were:

1. CGA TTC GGCG (91,299) MgCl$_2$ conc. 3 mM.
2. CGA GGT TCGC (91,300) MgCl$_2$ conc. 2 mM.

Methods

DNA extraction

1. *Heterobasidion* isolates were cultured for 10–15 days at 20°C on malt agar plates containing sterile cellophane membranes on the surface. The mycelium was scraped off and ground to a fine powder under liquid nitrogen with a mortar and pestle. Samples which are not extracted immediately are stored at −70°C or lyophilized and stored at 4°C. DNA extraction is performed essentially according to Lee and Taylor (1990) with small modifications.
2. Fill 1.5 ml Eppendorf tubes one-third to half full with ground mycelium (20–60 mg dry or 100–300 mg fresh).
3. Add 400–700 μl lysis buffer and 40–70 μl RNase A.
4. Incubate at 65°C for 1 h.
5. Add 400 μl phenol (if more lysis buffer was added use an equal amount of phenol). Vortex.
6. Microcentrifuge at 10,000 g for 15 min or until the top phase is clear.
7. Transfer aqueous phase into a new tube. Be careful not to take the cellular debris from the interface.

8. Repeat three previous steps first with phenol–chloroform (1:1) and then with chloroform.
9. Add 10 μl 3 M sodium acetate followed by 0.54 vol. of isopropanol. Leave to precipitate for 5 min.
10. Microcentrifuge 2–5 min as above. Pour off supernatant and rinse the pellet once with 70% ethanol. Invert the tubes for 1 min and drain them on a paper towel.
11. Leave tube caps open and dry pellets at 37°C for 15 min to 1 h.
12. Resuspend the pellet in 25–50 μl TE.
13. The final DNA concentration can easily be determined by using ethidium bromide plates (0.5% agarose, EtBr 2 μg ml^{-1}). Apply 0.5 μl samples of the DNA solution and duplicates which are diluted × 10 on to the plate. Also apply suitable standard DNA (λ-DNA) at 10–250 ng μl^{-1} concentrations.
14. Store overnight at 4°C and take a photo of the plate.
15. Determine the concentration by comparing samples with the standards.
16. Make dilutions to contain 10 ng DNA μl^{-1} in TE. Store template and dilutions at 4°C.

RAPD PCR amplification   We have amplified *Heterobasidion* DNA by PCR using the modified protocol of D. Howland and J. Arnau (University of East Anglia, Norwich, UK). Most of the work has been performed in a laminar flow cabinet.

1. Make a cocktail containing: 200 μM of each dNTP; 200 nM of primer; 1 × buffer containing a suitable concentration of MgCl$_2$; 1 U polymerase/each reaction (reaction volume 50 μl); sterile water to adjust the volume.
2. Divide this pre-mixture into 0.5 ml PCR tubes and add 25 ng of template-DNA (= 47.5 μl pre-mixture and 2.5 μl DNA-dilution).
3. Overlay each reaction with 2 drops of sterile paraffin oil.
4. Spin down briefly.

DNA amplification (40 cycles) was performed in an MJ programmable Thermal Cycler (MJ Research, Inc.) as follows:

1. 1 min 94°C; 1 min 45°C; 2 min 72°C.
2. Freeze reactions after amplification. Remove paraffin oil from the top of the frozen reaction either by pipetting or with a tip attached to vacuum.
3. Amplification products were analysed with ordinary agarose TBE gel (1–1.4%) stained with ethidium bromide.
4. The gels were photographed under UV light.

Data treatment

1. Resulting bands are detected visually and recorded as 1/0 matrix (presence/absence).
2. We have used the NTSYS-pc program (Rohlf, 1989) to analyse the band matrix but several other programs can also be used for RAPD data handling (Demeke *et al.*, 1992; Wilde *et al.*, 1992). Similarity coefficients, DICE, are calculated and phylogenetic trees constructed from that by SAHN clustering (UPGMA method).

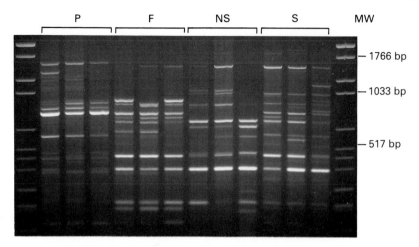

Fig 16.1. Amplification of genomic DNA from three *Heterobasidion* isolates of different intersterility groups. P, pine specific; F, fir specific; S, spruce specific; NS, North American S-type. Fragments were analysed on 1.4% agarose-tbe gel alongside plasmid marker pBR328 DNA *BgI* + *HinI*. The RAPD primer 91299, CGA TTC GGCG.

RESULTS

RAPD amplification of *Heterobasidion* DNA revealed different fingerprints for each major intersterility group (Fig. 16.1). The fragment pattern of the F group is closer to North American S and European S than to European P, suggesting that P-group isolates are genetically more distant. This pattern was consistently found with a number of different primers. Analysis of a large collection of P and S isolates clearly demonstrated that there is considerably more polymorphism in the S group than in the P group. Thus, our experience suggests that PCR amplification using RAPD markers is a rapid and convenient way for the identification of intersterility groups in *H. annosum*.

LIMITATIONS

The efficacy of PCR-RAPD analysis depends on various factors, but the optimal concentration of template DNA (about 25 ng), $MgCl_2$ concentration (2–3 mM) and annealing temperature (40°C) were the most important parameters for obtaining the best resolution. The mini-prep protocol of Lee and Taylor (1990), with slight modifications as described here, provided enough high quality DNA from *Heterobasidion* for successful amplification. However, the dilution of DNA with sterile water leads to unsuccessful amplification and it was necessary to dilute the template DNA with TE. The process of DNA dilution appears to be one major source of contamination which may easily cause false results. The level of $MgCl_2$ concentration also had a great impact on band resolution as has been previously reported (Williams *et al.*, 1990), and it was necessary to optimize the $MgCl_2$ level separately for each primer.

The complex patterns of amplified products caused difficulties in interpreting the results. In order to analyse the banding patterns, estimates of similarities were used to

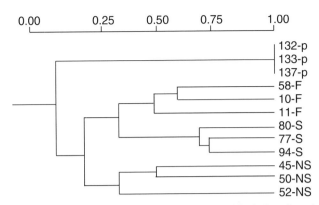

Fig 16.2. UPGMA dendrogram of relationships among *Heterobasidion* isolates based on similarity (DICE) coefficient using SAHN computer program (NTSYS-pc, Rohlf 1989). Similarity matrix based on the presence/absence of easily detectable bands.

generate a similarity matrix and hence a dendrogram (Fig. 16.2). There is no consensus of whether it is best to count all bands or to weigh them according to the intensity of bands. Recently, Demeke *et al.* (1992) scored RAPD bands into six classes on the basis of their size and intensity and found this more useful than the presence/absence scale (Fig. 16.2).

Clustering analysis clearly separates the major intersterility groups into distinct RAPD classes and also indicates that North American S is a separate group from the corresponding European S. Consequently, our experience suggests that isolates from *H. annosum* can be easily grouped by their RAPD patterns when the data are treated by clustering analysis.

The distinct groups found in RAPD analysis (Figs 16.1 and 16.2) suggest the usefulness of this method for revealing interesting phylogenetic patterns in fungi. For example, our data clearly suggest that there are three genetically isolated populations of this fungus in Europe. Two of these groups have a narrow host range: F group attacks mainly silver fir (*Abies alba*) and S Norway spruce (*Picea abies* L.), while P-group isolates can infect several tree species (Korhonen, 1978; Capretti *et al.*, 1990). However, it is interesting that within S group there was a higher amount of polymorphisms than in P group although P has a wider host range. Historical events related to differences in the spread of the host trees since the last ice age in Europe might explain the genetical differences in *Heterobasidion* more than the current host range of the isolates.

REFERENCES

Capretti, P., Korhonen, K., Mugnai, L. and Romagnoli, C. (1990) An intersterility group of *Heterobasidion annosum* specialized to *Abies alba. European Journal of Forest Pathology* 20, 231–240.

Chase, T. E. and Ullrich, R. C. (1988) *Heterobasidion annosum*, root- and butt-rot of trees. *Advances in Plant Pathology* 6, 501–510.

Demeke, T., Adams, R. P. and Chibbar, R. (1992) Potential taxonomic use of random amplified polymorphic DNA (RAPD): a case study in *Brassica. Theoretical and Applied Genetics* 84, 990–994.

Harrington, T. C., Worrall, J. J. and Rizzo, D. M. (1989) Compatibility among host-specialized isolates of *Heterobasidion annosum* from western North America. *Phytopathology* 79, 290–296.

Hodges, C. S. (1969) Modes of infection and spread of *Fomes annosus*. *Annual Reviews of Phytopathology* 7, 247–266.

Karlsson, J. O. and Stenlid, J. (1991) Pectic isoenzyme profiles of intersterility groups in *Heterobasidion annosum*. *Mycological Research* 95, 531–536.

Korhonen, K. (1978) Intersterility groups of *Heterobasidion annosum*. *Communications of the Institute of Forest Fenn.* 94, 1–25.

Lee, S. B. and Taylor, J. W. (1990) Isolation of DNA from fungal mycelia and single spores. In: Innis, M. A., Gelfand, D. H., Sninsky, J. J. and White, T. J. (eds) *PCR Protocol: A Guide to Methods and Applications*. Academic Press, San Diego, pp. 282–322.

Otrosina, W. J., Chase, T. E. and Cobb, F. W. (1992) Allozyme differentiation of intersterility groups of *Heterobasidion annosum* isolated from conifers in the western United States. *Phytopathology* 82, 540–545.

Rohlf, F. J. (1989) *NTSYS-pc. Numerical Taxonomy and Multivariate Analysis System*. Exeter Publisher, Ltd, Setauket, NY.

Smith, M. L., Bruhn, J. and Anderson, J. (1992) The fungus *Armillaria bulbosa* is among the largest and oldest living organisms. *Nature* 356, 428–456.

Wilde, J., Waugh, R. and Powell, W. (1992) Genetic fingerprinting of *Theobroma* clones using randomly amplified polymorphic DNA markers. *Theoretical and Applied Genetics* 83, 871–877.

Williams, J. G. K., Kubelik, A. R., Livak, K. J., Rafalski, J. A and Tingey, S. V. (1990) DNA polymorphisms amplified by arbitrary primers are useful as genetic markers. *Nucleic Acids Research* 18, 6531–6535.

17

The Development of Rapid Identification of *Heterobasidion annosum* Intersterility Groups by Ribosomal RNA Gene Polymorphism and by Minisatellite Allele Polymorphism

T. KASUGA AND K. R. MITCHELSON

Department of Molecular and Cell Biology, University of Aberdeen, Marischal College, Broad Street, Aberdeen AB9 1AS, UK.

BACKGROUND

Heterobasidion annosum is an important basidiomycete causing wood rot of forest trees. The fungus has a wide geographic distribution and infects a large range of species, principally conifers but also commonly other forest species, with varying pathogenicity. In 1978, Korhonen (1978) demonstrated two European intersterility groups (ISGs) based on incompatability *in vitro* which were termed 'Pine' (*Pinus* spp.) and 'Spruce' (*Picea* spp.) groups, respectively, after the major host of each group. More recently, a third group restricted to *Abies alba* and termed 'Fir' was identified by Capretti *et al.* (1990). Intersterility testing of European isolates has shown that the ISG-P are found on a number of conifers including *Pinus*, *Picea*, *Larix* and *Abies*, and commonly on broad-leaf species (Korhonen, 1978; Capretti, *et al.*, 1990; Stenlid and Karlsson, 1991). The ISG-S group prefers *Picea*, but may be found on *Abies* in northern Europe and the ISG-F is specialized to *Abies alba* in Italy and France.

The classification of the three European ISGs [and two groups, ISG-S and ISG-P, in North America (Chase and Ullrich, 1990)] have been made using classical, non-molecular methods such as *in vitro* intersterility, host preference, basidiocarp morphology, etc. These procedures have several drawbacks including the long times taken for intersterility reactions to develop (weeks/month), the potential inaccuracy of ISG determination due to partial interfertility between ISGs, and the variability of the interaction morphology, such as the failure of some compatible strains to form clamp connections. These factors combine to give variable reactions to some tests and also make intersterility tests difficult to interpret in all cases.

DNA-based techniques for the identification of *H. annosum* ISGs which have been developed include ribosomal DNA sequence polymorphism (Kasuga *et al.*, 1993; Karjalainen and Fabritius, 1993a), PCR amplification of random DNA sequences (RAPD) (Karjalainen and Fabritius, 1993; Karjalainen and Kammiovirta, Chapter 16) and minisatellite DNA polymorphisms. We report here the methods for detection of ribosomal DNA polymorphism and minisatellite polymorphisms to distinguish ISGs of *H. annosum*.

Ribosomal DNA polymorphisms

Ribosomal DNA (rDNA) repeat units contain highly conserved DNA sequences as well as more variable DNA sequence regions. The intragenic transcribed spacer (ITS) comprises the transcribed region flanking the 5.8S gene and has been used to detect genetic variation between related fungal species (White *et al.*, 1990). Amplification of the ITS region by the polymerase chain reaction (PCR) using DNA primers specific for conserved flanking 18S and 28S elements produces characteristic DNA fragments in *H. annosum* (Kasuga *et al.*, 1993). The DNA sequence of the ITS region was determined in the three European ISGs collected over a large region of northern and western Europe and was found to be constant for ISG-P and ISG-F isolates. Greater variability was found in ISG-S isolates (Fig. 17.1). Korhonen *et al.* (1992) have suggested that as a result of isolation of small *Picea abies* populations during the last ice age two ISG-S populations of *H. annosum* occur in Europe: a northern European (Scandanavia/Russia to central Europe) and a southern European/Alpine population. The differences in the ITS region of ISG-S may reflect local population variation or may result from interaction between populations of ISG-S. A detailed analysis of isolates from both populations using rDNA, minisatellite and other genetic markers in combination with intersterility testing should resolve this question. Restriction analysis of the ITS amplification product showed RFLPs could be detected between ISG-P isolates and either ISG-S/-F by at least three enzymes, as shown in Fig. 17.2, or a combination of enzymes (Kasuga *et al.*, 1993). The sequence differences detected between ISG-S variants and between ISG-S and ISG-F occur at sequence sites not amenable to restriction. Some discrimination between these two groups, either by differential PCR amplification or by single DNA strand conformational polymorphism (SSCP), was also developed (Table 17.1 and Fig. 17.3).

PROTOCOL: ITS AMPLIFICATION, RESTRICTION POLYMORPHISM AND SSCP ANALYSIS

DNA preparation

Materials

1. Porcelain mortar and pestle, liquid nitrogen, paper towels.
2. Extraction buffer (Richards, 1987) containing 100 mM Tris-HCl (pH 8), 100 mM EDTA, 250 mM NaCl and 100 μg ml^{-1} proteinase K. Also 10% sodium sarcosyl, isopropanol, TE buffers and 10 mg ml^{-1} ethidium bromide.

ISG-S:			ITS1		5.8S		ITS2	
S$_{Fi1}$	GG	C	TTGCG C	*	####	C C T	**	A G T A C
Ha37				*	####	C	**	A
Ha47			T	*	####	C	**	*
Ha35				*	####	C	**	*
S$_{Am}$	A	T		*	####	T C	**	G C *
ISG-F:				*	####	C	**	*
ISG-P: AA		T	GAC** T	*	####	T C		TG G A A T

Fig 17.1. Diagrammatic representation of nucleotide variation in the ITS sequence of the ribosomal gene repeat between ISGs. Variant nucleotides occur in the ITS1 and ITS2 regions at the positions indicated. Ha35 has an identical ITS sequence to ISG-F but was identified as ISG-S by intersterility test and by minisatellite M13 probe. S$_{Am}$ is North American ISG-S, and is shown for comparison. #### = 5.8*S* gene region.

Methods

1. Ten day mycelial cultures were grown on liquid media (Norkrans, 1963). Mycelia were washed in distilled, autoclaved H$_2$O to remove all growth media, then dried on paper towels and weighed.

2. Mycelia were ground under liquid N$_2$ to a fine powder. For small-scale preparations for PCR less than 1 g of mycelia was used.

3. For large scale DNA preparations, 10 g or more of powdered mycelia were allowed to thaw in 15 wt. vols of Richards's extraction buffer.

4. Sodium sarcosyl was added to 1% and the lysate incubated at 55°C for 2 h. (Solutions should be handled gently from this stage to prevent shear of DNA. Do not vortex or mix vigorously.)

5. Cell debris was removed by centrifugation at 7000 rpm 15 min at 4°C in a Beckman HS-4 rotor. The supernatant was filtered through gauze to remove any particulate matter.

6. DNA was precipitated by addition of 0.6 vol. of isopropanol, incubation at −20°C for 30 min and pelleted by centrifuging at 8000 rpm in a Beckman GS-3 rotor for 15 min at 4°C. The crude DNA was resuspended in 1×TE buffer.

Enzyme	*Bst*U1	*Hha*I	*Mbo*I
ISG			
European			
ISG-P	254,**193**,113,38,35	331,228,74	230,212,191
ISG-S	254,113,107,**84**,40,35	226,**193,140**,57,17	232,212,189
ISG-F	254,113,106,**84**,40,35	225,**193,140**,57,17	232,212,188
Other			
P$_{Am}$	254,113,106,83,40,35	331,226,74	
S$_{Am}$	254,119,113,106,40	225,193,140,57,17	
Hara	378,116,107,45	342,194,110	

Fig 17.2. Polymorphic restriction fragments (in bp) of the ribosomal *H. annosum* ITS region PCR amplified by ITS1 and ITS4 primers (White *et al.*, 1990).'Other' refers to *H. annosum* from non-European lands: shown are the restriction fragments of the ITS region of North American ISG-P (P$_{Am}$) and ISG-S (S$_{Am}$) and also *Heterobasidion araucariae* Hara from Australia.

Table 17.1 Intersterility group determined by independent tests, collector and reference number.

Isolate	ITS	MS	ISG	Col	Ref	Local/number	
ISG-P							
GB1	P	P	P	JG	—	Eng	Oxford
GB3	P	P	—	CW	2,3	Sco	Glen Muick
GB4	P	P	—	CW	2,3	Sco	Glen Muick
GB6	P	P	—	DR	2,3	Sco	Peebles
Ha4	P	—	P*	CD	2,3	Fra	Neufchateau
Ha45	P	P	P*	CD	2,3	Fra	Neufchateau
Ha78	P	—	P*	CD	2,3	Fra	Neufchateau
Ha27	P	P	P	OH	—	Swi	871214.1.1/2
ISG-S							
S_{Fi1}	S	S	S	KK	1,2	Fin	870923.3.1/4
Ha64@	S+F	—	S*	CD	3	Fra	Neufchateau
Ha55	F	—	S*	CD	3	Fra	Neufchateau
Ha37	S	—	S	PC	1,2	Ita	871104.3.1./5
Ha35	F	S	S	PC	1,2	Ita	871104.1.1/1
Ha44	S	S	S	KK	2	Bye	900913.4.11/4
Ha47	S	—	S	KK	2	Fin	911108.1.5/1
ISG-F							
F_{Ha1}	F	F	F	PC	1,2	Ita	870900.1.2/4
F_{Ha5}	F	F	F	PC	1,2	Ita	871022.2.3/1
F_{Ha8}	F	F	F	PC	1,2	Ita	871022.5.1/6
Non-European							
P_{NAm}	P		P	TC		USA	880120.1.1
S_{NAm}	P		P	TC		USA	880120.1.9/1
Hara	O		O	LB		Aust	651100.1.3

Abbreviations: ITS: intragenic transcribed spacer and 5.8S rRNA gene sequence or SSCP test; **MS**: minisatellite loci; **ISG**: intersterility group determined by sexual incompatibility *in vitro*, reported in given references.
Col: collector. **Ref**: reference, details of isolation. **Local**: Bye (Byelorussia); Eng (England); Fin (Finland); Fra (France); Ita (Italy); Sco (Scotland), Swi (Switzerland), USA (North America), Aust (Australia). *Heterokaryon classified against Finnish tester strains. @Ha64 is a dikaryotic culture with ITS characteristics of both ISG-S and ISG-F.
References: (1) Capretti *et al.* (1990) (2) Korhonen *et al.* (1992) (3) Kasuga *et al.* (1993).
Collectors: LB: L Bolland, PC: Paulo Capretti, TC: Tom Chase, CD: Claude Delatour, JG: John Gibbs, OH: Ottmar Holdenreider, KK: Kari Korhonen, DR: Derek Redfern, CW: Caroline Woods.

7. DNA was further purified by CsCl isopycnic centrifugation in the presence of 0.3 mg ml^{-1} ethidium bromide (Maniatis *et al.*, 1982). This last step was not necessary for PCR purposes.

ITS amplification and SSCP

Materials

1. Techne PHC-3 or Hybaid thermal cyclers.
2. 15×15 cm polyacrylamide gel electrophoresis system (Hoeffer). 1 × TBE buffer

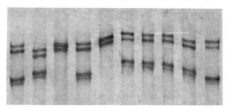

Fig 17.3. Detection of SSCP in PCR generated ITS3–4 region of *H. annosum*. The polyacrylamide gel contained 0–10% glycerol (see methods). Fungal isolates were from different European locations. Lanes were P_{GB1} (1), P_{NAm} (2), S_{Fi1} (3), S_{Ha47} (4), S_{Ha44} (5), S_{Ha55} (6), S_{Ha35} (7), S_{Ha37} (8), F_{It} (9), and S_{NAm} (0).

(Maniatis *et al.*, 1982). DNA loading buffer for electrophoresis contained 20 mM EDTA (pH 8), 95% formamide, 0.05% bromophenol blue and 0.05% xylene cyanol.

3. Reagents for silver staining (Bassam *et al.*, 1991), 25×25 cm plastic dish, water pump, chromatography paper, vacuum gel drier.

Methods

1. The ITS region was amplified in a reaction containing 0.225 μM of the universal fungal rDNA primers ITS1 and ITS4 (White *et al.*, 1990).
2. The amplification products were digested by single restriction enzymes: *Hha*I, *Bsu*UI, *Mbo*I, *Msp*I, *Hinf*I, or by combinations of enzymes: *Hinf*I plus *Hae*III. Some restriction products are listed in Fig. 17.2.
3. The digestion products were co-electrophoresed in a 7% non-denaturing polyacrylamide gel using 1×TBE buffer.

Silver staining of DNA

1. DNA fragments were visualized by silver staining (Bassam *et al.*, 1991). The polyacrylamide gel was removed from the glass electrophoresis plates and immersed in a series of fixing, washing and staining solutions in a plastic dish. A water pump was used to remove each solution leaving the gel in the dish.
2. Fix DNA gel for 20 min in 10% acetic acid with gentle agitation throughout. Wash the gel three times for 5 min each with large vol. dH$_2$O with gentle agitation.
3. Stain gel for 30 min in 1 g l^{-1} AgNO$_3$ containing 1.5 ml l^{-1} of 40% formaldehyde.
4. Develop the gel for 5 min in 1 l of 30 g l^{-1} Na$_2$CO$_3$ containing 1.5 ml l^{-1} of 40% formaldehyde and 0.1 g of sodium thiosulphate.
5. The rate of development of the silver stain is reduced by increasing the amount of sodium thiosulphate. The reaction must be closely monitored and stopped as soon as DNA bands are clearly visible. Overdevelopment is prevented by addition of 20% acetic acid directly to the development solution to 10% final concentration.
6. The stained gel was placed on chromatography paper and vaccum dried.

Differential PCR amplification

1. Specific oligonucleotide primers were designed to discriminate between ISGs. The assay was based on 3' match of the primer on correct DNA templates which would allow amplification of a PCR product, whereas mismatched primer/templates would prevent amplification.

2. Two oligonucleotide primers were designed for ISG-P and ISG-F, respectively. ISG-P: 5'-CCCACAATCGTGGCGTACCA, used in conjunction with ITS3 primer produces a product of 291 bp with ISG-P. Reaction conditions: 94°C, 1 min; 60°C, 1 min; 72°C, 1 min for 30 cycles then 5 min extension at 72°C. ISG-PS: 5'-GTCAAGTTTCGATAAAGTTT, used in conjunction with ITS3 primer produces a product of 337 bp in ISG-P and most ISG-S. Reaction conditions: 94°C, 1 min; 60°C, 1 min; 72°C, 1 min for 30 cycles then 5 min extension at 72°C.

3. Although this test worked reproducibly in our hands and was confirmed by DNA sequence analysis, a positive test was also designed to discriminate between ISG-F and ISG-S.

Single DNA strand conformational polymorphism (SSCP)

1. Most variation between ISGs was found in the ITS2 region. PCR products of ISG-S and ISG-F were synthesized using ITS3 and ITS4 primers as described above. The PCR products ($3 \mu l$) were mixed with loading buffer ($7 \mu l$) and then were denatured by boiling for 3 min and snap frozen before electrophoresis on 7% non-denaturing polyacrylamide gels (acrylamide : bis-acrylamide = 89 : 1) containing varying amounts of glycerol (0–10%) at 4°C and 12 watts constant power for 3–5 h.

2. DNA strands were visualized by silver staining, as above (Fig. 17.3). SSCP was seen between ISG-P and the other ISGs. Variants of ISG-S such as Russian/Finnish ISG-S (S_{Ha44} and S_{Fi1}) were distinguishable, other ISG-S variants and ISG-F. The SSCP in S_{Ka44} is due to a single $C \rightarrow T$ transition.

Minisatellite DNA polymorphisms

'Minisatellite' elements in man consist of a 16–36 bp element which is organized in tandem arrays throughout the genome; the arrays are polymorphic in length (Jeffreys *et al.*, 1985). Human minisatellites 33.15 and 33.6 and the repeat DNA element from the bacteriophage M13 (Vassart *et al.*, 1987) may be used to distinguish the European ISGs of *H. annosum*. Each minisatellite probe can detect different polymorphic DNA elements in *H. annosum*. Each probe also shows that geographically distant isolates from the *same* ISG have more similar fingerprints than other ISGs, and ISG-specific bands were detected. ISG-F and ISG-S shared more bands than with ISG-P, but discrete bands were detected in both ISG-F and ISG-S isolates. This suggests that DNA fingerprinting may aid discrimination of ISG-F and ISG-S.

Materials

1. Agarose (low EEO) (Sigma Chemicals), Loening's Buffer E (Maniatis *et al.*, 1982). Flat bed 20×20 cm electrophoresis tank (Pharmacia).

2. Polyurethane foam (10 cm thick), 2×SSC, Hybond-N membrane (Amersham).

Methods

DNA isolation and digestion by restriction endonucleases

1. DNA was isolated according to Richards (1987).
2. Southern blots of $5\,\mu g$ of genomic DNA could be digested using 4 or 5 bp cutters such as *AluI*, *HinfI*, *MspI*, *Sau3AI* or *MboI* to generate useful fingerprints. Each restriction enzyme produced complex fingerprint patterns, typically showing 20 or more strongly and weakly hybridizing bands.
3. *HaeIII* digests were not as useful as bands were less numerous and were smaller. Digestion with *EcoRI*, *HindIII* or *PstI* produced large DNA fragments. These were less easily resolved and had hybridization patterns of low complexity.
4. Multiple polymorphic DNA bands were detected when the blot was hybridized with M13 repeat (Fig. 17.4) or with minisatellites 33.15 and 33.6.

Electrophoresis and DNA hybridization conditions

1. The DNA fragments were restricted (see below) and electrophoresed on $20\,cm^2$, horizontal, 1.4% agarose gels using recirculated Buffer E at $1\,V\,cm^{-1}$ for 16 h.
2. The DNA was depurinated (10 min), denatured and transferred by Southern blotting on to Hybond-N membrane by supporting the agarose gel on filter paper on a pad of polyurethane foam immersed in 2×SSC. The foam ensures high rates of transfer to a stack of paper towels above the gel.

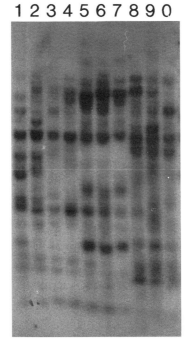

1 2 3 4 5 6 7 8 9 0

Fig 17.4. M13 DNA fingerprint of *HinfI* restriction of (1) S_{Fi1}, (2), S_{Ha35}, (3) S_{Ha64}, (4) S_{Ha44}, (5) F_{Ha1}, (6) F_{Ha5}, (7) F_{Ha8}, (8)P_{GB1}, (9) P_{Ha45}, (0) $P_H{}^a{}_{27}$. M13 repeat element detects prominent common bands within European ISGs. Several bands are seen in each ISG that are not detected in the other two groups.

3. The DNA was bonded to the membrane by baking at 80°C for 2 h followed by irradiation by UV light (312 nm, 20 s).

Bacteriophage M13 repeat probe   Prehybridization and hybridization with a 203 bp M13 element was performed for 24 h at 60°C in 7% SDS, 0.27 M sodium phosphate (pH 7.0), 1 mM EDTA, 1% BSA (Sigma Chemicals) (modified from Church and Gilbert, 1984). Membranes were washed at 56°C in 1×SSC and 0.1% SDS.

33.15 and 33.6 minisatellite probes

1. Cloned minisatellite elements were labelled with $[a-^{32}P]dCTP$ using random oligonucleotide primers (Feinberg and Vogelstein, 1983).
2. Prehybridization, hybridization and washing was carried out according to Daly *et al.*, (1991) with minor modifications. Briefly, hybridization was undertaken in the presence of $5 \times SSC$, 50% formamide, 1% skimmed milk powder, 25 mM sodium phosphate (pH 6.8) and 10% dextran sulphate for 18–30 h at 42°C.
3. Unbound probe was removed by washing at 60°C in $2 \times SSC$ (2 washes of 15 min), followed by a single 30 min wash in $0.1 \times SSC$ at 60°C.
4. Membranes were exposed to X-ray film (Fuji, RX) for 1–6 days at −70°C.

INTERPRETATION OF RESULTS

1. The relatedness of isolates was calculated by the index of similarity method of Nei and Li (1979). Restriction fragments detected by minisatellites that are common between each pair of isolates were counted for each enzyme and were used to determine the proportion of common fragments.
2. Minisatellites 33.15, 33.6 and M13-repeat detected different families of fragments. Each fingerprint may be used independently to calculate an index of similarity, or the results of several restriction digests or different minisatellite probes may be combined. ISG-specific bands were detected using M13 repeat probe (Fig. 17.4) suggesting that DNA fingerprinting may allow determination of ISG-S and ISG-F isolates.
3. A limitation on the technique is the variation in the intensity of hybridizing bands and the high number of bands apparently 'in common' between the ISGs.
4. Fingerprinting using all three probes provides independent data of repeat families. Data may be combined in analysis of an index of similarity.
5. Isolation of homologous polymorphic DNA fragments from *H. annosum* may provide more ISG-specific signals.

LIMITATIONS

1. The limitation on the discrimination between ISG-S and ISG-F by ITS sequence variation is apparent: (i) the ITS sequences several ISG-S isolates which are possibly identical to ISG-F (Fig. 17.1 and Table 17.1); (ii) the variation in ITS sequence between ISG-F and some ISG-S isolates occurs at nucleotides that are not amenable

to restriction; (iii) the variation in the ITS sequence within ISG-S may prevent easy identification of ISG-S although SSCPs have been detected in several ISG-S variants (Fig. 17.3).

2. Tests are underway to determine whether additional nucleotide variants of ISG-S and ISG-F may be resolved by SSCP following restriction of ITS2 region.

ACKNOWLEDGEMENTS

This work was supported by a research grant from the University of Aberdeen and by the UFC-Biotechnology Initiative and the NERC to KM. TK would like to acknowledge support of Mr H. Kasuga. The authors would also like to thank K. Korhonen, C. Wood and C. Delatour for fungal isolates and L. Allen and A. Cummings for advice on DNA conformational polymorphisms.

REFERENCES

Bassam, B. J., Caetano-Anolles, G. and Gresshoff, P. M. (1991) Fast and sensitive silver staining of DNA in polyacrylamide gels. *Analytical Biochemistry* 196, 80–83.

Capretti, P., Korhonen, K., Mugnai. L. and Romnagoli, C. (1990) An intersterility group of *Heterobasidion annosum* specialised to *Abies alba*. *European Journal of Forest Pathology* 20, 233–240

Chase, T. E. and Ulrich, R. C. (1990) Genetic basis of biological species in *Heterobasidion annosum*. Mendelian determinants. *Mycologia* 82, 67–72

Church, G. M. and Gilbert, W. (1984) Genomic sequencing. *Proceedings of the National Academy of Sciences of the United States of America*, 81, 1991–1995.

Daly, A., Kellam, P., Berry, S. T., Chojecki, A. J. S. and Barnes, S. R. (1991) The isolation and characterisation of plant sequences homologous to human hypervariable minisatellites. In: Burke, T., Dolf, G., Jeffreys, A. J. and Wolff, R. (eds) *DNA Fingerprinting: Approaches and Applications*. Birkhauser Verlag, Basel, Switzerland, pp. 330–341.

Fabritius, A. L. and Karjalainen, R. (1993) Variation in *Heterobasidion annosum* detected by random amplified polymorphic DNAs. *European Journal of Forest Pathology* 23, 193–200.

Feinberg, A. P. and Vogelstein, B. (1983) A technique for radiolabelling DNA restriction endonuclease fragments to high specific activity. *Analytical Biochemistry* 132, 6–13.

Jeffreys, A. J., Wilson, V. and Thein, S. L. (1985) Hypervariable 'minisatellite' regions in human DNA. *Nature* 314, 67–73.

Karjalainen, R. and Fabritius, A-L. (1993) Identification of intersterility groups of *Heterobasidion annosum* by restriction analysis of PCR products. *Journal of Phytopathology* (in press).

Kasuga, T., Woods, C., Woodward, S. and Mitchelson, K. R. (1993) *Heterobasidion annosum* 5.8S ribosomal DNA and internal transcribed spacer sequence: rapid identification of European intersterility groups by ribosomal DNA restriction polymorphism. *Current Genetics* 24, 433–436.

Korhonen, K. (1978) Intersterility groups of *Heterobasidion annosum*. *Communications of the Institute of Forest Fenn*. 94, 1–25

Korhonen, K., Bobko, I., Hanso, S., Piri, T. and Vasiliauskas, A. (1992) Intersterility groups of *Heterobasidion annosum* in some spruce and pine stands in Byelorussia, Lithuania and Estonia. *European Journal of Forest Pathology* 22, 384–391.

Maniatis, T., Fritsch, E. F. and Sambrook, J. (1982) In: *Molecular Cloning: A Laboratory Manual*. Cold Spring Harbour Press, Cold Spring Harbour, New York, USA.

Nei, M. and Li, W. H. (1979) Mathematical model for studying genetic variation in terms of restriction endonucleases. *Proceedings of the National Academy of Sciences of the United States of America* 76, 5269–5273.

Norkrans, B. (1963) Influence of some cultural conditions on fungal cellulose production. *Physiologica Plantarum* 16, 11–19

Richards, E. (1987) Preparation of genomic DNA from plant tissue. In: Ausubel, F. M. (ed.) *Current Protocols in Molecular Biology.* Greene Publishing Associates, New York, USA, pp. 2.3.1–2.3.3.

Stenlid, J. (1985) Population structure of *Heterobasidion annosum* as determined by somatic incompatability, sexual incompatability and isoenzyme patterns. *Canadian Journal of Botany* 63, 2268–2273.

Stenlid, J. and Karlsson. J. O. (1991) Partial intersterility in *Heterobasidion annosum. Mycological Research* 95, 1153–1159.

Vassart, G., Georges, M., Monsieur, R., Brocas, H., Lequarre, A. S. and Christophe, D. (1987) A sequence in M13 phage detects hypervariable minisatellites in human and animal DNA. *Science* 235, 683–684.

White, T. J., Bruns, T., Lee, S. and Taylor, J. (1990) Amplification and direct sequencing of fungal ribosomal RNA genes for phylogenetics. In: Innes, M. A., Gelfrand, D. H., Sninsky, J. J. and White, T. J. (eds) *PCR Protocols: A Guide to Methods and Applications.* Academic Press, San Diego, USA, pp. 315–322.

18 Identification of Fungi in the *Gaeumannomyces–Phialophora* Complex Associated with Take-all of Cereals and Grasses Using DNA Probes

E. WARD AND G. L. BATEMAN

AFRC Institute of Arable Crops Research, Rothamsted Experimental Station, Harpenden, Herts AL5 2JQ, UK.

BACKGROUND

The pathogens that cause take-all are part of the *Gaeumannomyces–Phialophora* complex of fungi that infect the roots of graminaceous plants. *G. graminis* var. *tritici* (Ggt) causes take-all in cereal crops, mostly wheat and barley. *G. graminis* var. *avenae* (Gga) causes take-all in oats, but is found only in the western parts of the British Isles where more oats are grown, even though it is pathogenic on other cereals; it occurs more widely as the cause of take-all patch in turf grasses. These fungi have *Phialophora*-like anamorphs. Avirulent and weakly pathogenic *Phialophora* species also occur on cereal and grass roots. Recognized species are *Phialophora* sp. (lobed hyphopodia), some isolates of which produce perithecia of *G. graminis* var. *graminis* (Ggg) (Walker, 1981; Hornby *et al.*, 1991), and *P. graminicola* (Deacon) Walker (teleomorph: *G. cylindrosporus* D. Hornby *et al.*). These are of particular interest in epidemiological research because of their ability to protect against infection by the take-all fungi. The taxonomy of the group has been discussed by Walker (1981).

Diagnosis of take-all may be difficult where other fungi, e.g. *Pythium* and *Fusarium* spp., also cause root discoloration. The pathogenic and non-pathogenic members of the *Gaeumannomyces–Phialophora* complex all produce dark ectotrophic hyphae on roots. Some closely resemble each other in culture and all grow on a selective medium (G. L. Bateman, unpublished) developed by Juhnke *et al.* (1984) that is used to confirm diagnoses of take-all in parts of the USA where the avirulent *Phialophora* species are apparently absent. Confirming diagnoses of fungi isolated from diseased plants conventionally by pathogenicity testing and by incubation to produce perithecia is laborious and time consuming and can be inconclusive because some isolates fail to produce perithecia (Holden and Hornby, 1981).

Molecular techniques have therefore been investigated as a way of finding a reliable means of confirming the identification of the take-all fungi and related fungi. The two DNA probes described here represent the first completed stage of a programme of work to develop practicable diagnostic procedures for use in take-all research.

PROTOCOL

Materials

Equipment   Incubators or water baths; freeze drier or source of liquid nitrogen; microcentrifuge; gel electrophoresis equipment; UV transilluminator and camera; vacuum blotting apparatus (optional); UV spectrophotometer (optional); thermal cycler (if generating probe by PCR); hybridization oven (optional); lightproof cassettes; X-ray developing machine (optional).

Other materials   Filter paper (Whatman 3 MM); nylon membranes (Hybond N, Amersham); DNA labelling and detection kits (Boehringer Mannheim); hybridization bags (Gibco-BRL); X-ray film (Fuji RX).

Probes

1. pMSU315. This is a 4.3 kb fragment of mitochondrial DNA from an isolate of Ggt inserted into the *Acc*I site of plasmid pUC18. It was generated by J. M. Henson, Department of Microbiology, Montana State University, Bozeman, Montana, USA who showed it to have homology with DNA from most isolates of fungi in the *Gaeumannomyces–Phialophora* complex from cereals and grasses (Henson, 1989, 1992). Its use in differentiating the pathogenic varieties from the non-pathogens was described by Bateman *et al.* (1992). It has been deposited with the European Collection of Animal Cell Cultures, Porton Down, Salisbury, Wiltshire, SP4 0JG, from where it is available.
2. GggMR1/pEG34. GggMR1 is a PCR-amplified mitochondrial ribosomal DNA fragment from an isolate of Ggg (89/5–1), generated and described by Ward and Gray (1992). This fragment was subsequently cloned into the *Eco*RV site of pbluescript SK+ using the protocol of Marchuk *et al.* (1991) and the resulting plasmid, pEG34, is available from E. Ward. It is used in the diagnosis of individual species and varieties in the *Gaeumannomyces–Phialophora* complex.

Buffers

1. L-broth (culture medium) – $10 \, g \, l^{-1}$ tryptone, $5 \, g \, l^{-1}$ yeast extract, $10 \, g \, l^{-1}$ NaCl.
2. Lysis buffer (DNA extraction) – $50 \, mM$ Tris pH 7.4, $50 \, mM$ EDTA, 3% SDS, 1% 2-mercaptoethanol.
3. TE – $10 \, mM$ Tris pH 7.5, $0.1 \, mM$ EDTA.
4. TBE – $50 \, mM$ Tris, $50 \, mM$ boric acid, $1 \, mM$ EDTA or TAE – $40 \, mM$ Tris, $10.2 \, ml \, l^{-1}$ glacial acetic acid, $1 \, mM$ EDTA.

5. PCR reaction buffer (× 10) – 200 mM Tris pH 8, 15 mM MgCl$_2$, 250 mM KCl.
6. Primers (for PCR). These were devised by White *et al.* (1990): MS1, CAGCAGTCAA-GAATATTAGTCAATG; MS2, GCGGATTATCGAATTAAATAAC.
7. SSC (20 × stock) for blotting and DNA probing – 3M NaCl, 0.3 M trisodium citrate, pH 7.
8. Hybridization solution (DNA probing) – 5 × SSC, 0.5% blocking agent (Boehringer), 0.1% N-laurylsarcosine, 0.02% SDS.
9. SSPE (20 × stock) – 3.6 M NaCl, 0.2 M NaH$_2$PO$_4$, 0.02 M EDTA, pH 7.7.

Methods

DNA extraction   The method is essentially that of Lee and Taylor (1990).

1. Incubate pure fungal cultures (15 ml L-broth) on an orbital shaker at room temperature (15–25°C) for 4–7 days.
2. Take out the mycelium, blot to remove most of the liquid and either:
   (a) transfer to a 2 ml screw-cap microcentrifuge tube in 0.3 g aliquots, freeze dry, and grind to a powder using a metal rod shaped to fit the tubes; or
   (b) grind with liquid nitrogen using a mortar and pestle. This is more suitable for larger volumes of mycelium and in this case the excess broth should be removed by vacuum filtration. Transfer to 2 ml screw-cap microcentrifuge tubes in 0.3 g aliquots.
3. Add 0.4 ml lysis buffer per tube, vortex and incubate at 65°C for 1 h.
4. Add 0.4 ml phenol:chloroform (1:1 w/v), vortex and centrifuge for 15 min.
5. Transfer 0.3–0.35 ml from the top (aqueous) layer into a clean tube. Repeat stages 4 and 5 if the aqueous phase is cloudy or contains debris.
6. Add 18 μl 6 M ammonium acetate (or 10 μl 3 M sodium acetate) and 0.54 volumes propan-2-ol to the top layer. Invert several times to mix and centrifuge for 2–5 min.
7. Pour off the supernatant, add 1 ml cold 70% ethanol, spin briefly and discard the supernatant.
8. Spin the tube briefly to collect excess liquid and remove this using a pipette.
9. Allow the remaining solvent to evaporate in a fume hood for 25 min or by using a vacuum centrifuge.
10. Resuspend the pellet in 50–200 μl TE and determine the concentration of DNA by measuring the A$_{260}$ on a UV spectrophotometer or by agarose gel electrophoresis and comparison with a standard of known concentration.

An additional RNase digestion (followed by phenol extraction and propan-2-ol precipitation) can be added to the protocol to remove contaminating RNA, but this is not essential.

Restriction enzyme digestion, electrophoresis and blotting

1. Digest the DNA (1–2 μg) for 3–18 h with *Bam*HI (for probe 1) or *Eco*RI (for probe 2) using a final volume of 10 μl and the buffer supplied/recommended by the manufacturer.
2. Electrophorese on a 1% agarose gel (with TAE or TBE electrolyte) containing 0.5 ng μl^{-1} ethidium bromide. Include a size marker such as λ-DNA cut with *Eco*RI and *Hind*III.

3. Photograph the gel in UV light and transfer the DNA to a nylon membrane by standard capillary blotting procedures (see Sambrook *et al.*, 1989) or using a vacuum blotting apparatus.

4. Fix the DNA to the filter by UV treatment or by baking it at 80°C for 3 h.

PCR   Probe GggMR1 was generated originally by PCR as described below using primers that recognize the ribosomal DNA of any fungus. Now that GggMR1 has been cloned we recommend that, in future, the resulting plasmid (pEG34) is used rather than the PCR-generated probe. However, the protocol is included here since it can be used to make similar probes for other fungi. Other primer pairs (White *et al.*, 1990) can also be used to make probes that recognize other regions of the ribosomal DNA.

1. Mix the following: distilled water (to give a final volume of 100 μl), 10 μl 10 × reaction buffer, 16 μl of dNTP mixture (1.25 mM of each of dATP, dCTP, dGTP and dTTP), 100 pmol each primer (MS1+MS2), 2.5 units *Taq* polymerase and 0.2μg fungal DNA.

2. Cover with 60 μl light mineral oil and place in a thermal cycler. The PCR conditions used are 25 cycles of 94°C for 1 min, 42°C for 2 min and 74°C for 3 min.

3. Remove the oil and check the product by electrophoresing on a 1% agarose gel.

DNA probing   We now routinely use the non-radioactive digoxigenin (DIG) DNA labelling and detection system from Boehringer Mannheim who supply full protocols and most of the ingredients for the method. Brief details are given below. Radioactive labelling procedures can also be used and the protocols for this are described in Bateman *et al.* (1992).

1. Label the probe (10 ng–3 μg DNA) overnight at 37°C in a final volume of 20 μl using the reagents and methods for random primer labelling in the kit. Plasmid DNA must first be linearized with a restriction enzyme (*Eco*RI is suitable for both pMSU315 and pEG34) and cleaned by phenol extraction and precipitation with propan-2-ol before labelling. For PCR-generated probes the products of two PCR reactions are combined and precipitated with propan-2-ol before labelling.

2. Stop the reaction and precipitate the DNA by adding 2 μl 0.2 M EDTA, 2 μl 4 M LiCl and 60 μl chilled ethanol. Resuspend the pellet in 50 μl distilled water. DIG-labelled probes can be stored at −20°C for many months.

3. Incubate the nylon filter at 68°C for 3–6 h in hybridization solution.

4. Denature the probe (95°C, 10 min) and add to 2 ml hybridization solution to give a final concentration of about 100 ng ml^{-1} (generally about 5 μl of the labelled probe is needed). Incubate the filter with this in a hybridization bag overnight at 68°C. A hybridization oven is better if one is available.

5. Wash the filters (high stringency conditions) twice for 5 min at room temperature in 2 × SCC, 0.1% SDS, followed by two washes for 15 min at 68°C in 0.1 × SSC, 0.1% SDS. Lower stringency can also be used for pMSU315; we normally wash twice for 5 min at 42°C in 0.5–2 × SSPE, 0.1% SDS.

6. The remaining stages of the DIG detection are as described in the manufacturer's instructions. Two detection methods are available, a colour-based system and chemiluminescent detection. Both give good results but the chemiluminescent method is preferable since it is more sensitive and the filters can be stripped and re-probed more

easily. With the chemiluminescent method, bands appear after exposure to X-ray film for 5 min to 24 h. The sizes of the labelled DNA bands are determined by comparing their positions with those of the size marker on the gel.

INTERPRETATION OF RESULTS

Probe pMSU 315

Isolates of Ggt and Gga produce a prominent, diagnostic band of about 2.9 kb (Fig. 18.1). Up to three other bands in the range 6.6–13.8 kb are usually produced for different isolates. Isolates of Ggg and *Phialophora* spp. usually produce a single band >20 kb (similar to a band sometimes produced for Ggt isolates). In tests of fungi outside the complex, no hybridization has been detected.

Probe GggMR1/pEG34

Isolates of Ggt produce bands of sizes 4.0 kb and 1.8 kb (type T1) or 5.4 kb and 1.8 kb (type T2) (Fig. 18.2). Combinations of these three different bands are considered diagnostic of Ggt; recently, some mass mycelial isolates have shown all three bands and may be mixtures of the types T1 and T2. A third band at 2.3 kb is also present for some isolates of Ggt.

Isolates of Gga all produce a band of 2.5 kb and a second band of either 5.4 kb (type A1) or 1.8 kb (type A2) (Fig. 18.2).

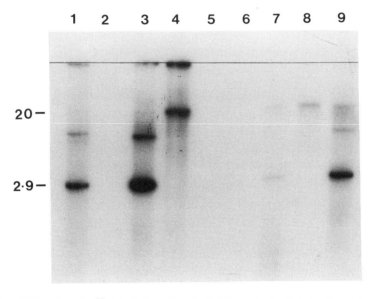

Fig 18.1. Hybridization of α-^{32}P labelled pMSU 315 with total DNA, digested with *Bam*HI, from isolates of *Gaeumannomyces graminis* var. *tritici* (lanes 1 and 3), *G. graminis* var. *avenae* (lanes 7 and 9), *G. graminis* var. *graminis* (lane 4) and *Phialophora* sp. (lobed hyphopodia) (lane 8). Other lanes (2, 5 and 6) contain DNA from other, unidentified, fungi from cereal roots.

Ggt   **Gga**   **Ggg**

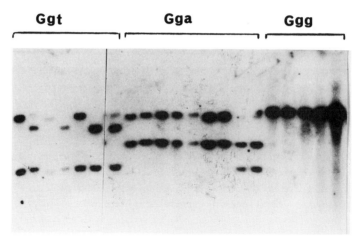

Fig 18.2.   Hybridization of digoxigenin-labelled GggMR1 with total DNA, digested with *Eco*RI, from isolates of *Gaeumannomyces graminis* var. *tritici* (Ggt), *G. graminis* var. *avenae* (Gga) and *G. graminis* var. *graminis* or *Phialophora* sp. (lobed hyphopodia) (Ggg). Detection was by chemiluminescence.

    Most Ggg and *Phialophora* sp. (lobed hyphopodia) isolates produce a single band of about 6.4 kb (type G1) (Fig. 18.2). Two isolates from a sedge, *Carex rostrata*, provisionally identified as Ggg, differed from this and also produced patterns different from other Gggs with pMSU 315 (Bateman *et al.*, 1992), as did an aggressive strain isolated from wheat in Poland. These may represent different, but closely related, taxonomic groups. Most *P. graminicola* isolates produce a single band of 2.5 kb (type P1) (not shown), but DNA of two isolates did not hybridize to the probe. Again, these may represent a different taxonomic group, or be aberrant forms of *P. graminicola*.
    DNA from fungi of unrelated taxa showed no or very little hybridization to the probe under the stringent conditions described.

USES AND LIMITATIONS

    Both these probing procedures are currently operational in our laboratory in diagnosis of the take-all pathogens, Ggt and Gga, and of the avirulent *Phialophora* spp. found on cereal roots. They have been used for early identification of new field isolates, especially ones with unusual appearance, and have helped in resolving identification problems and discrepancies in our existing culture collection. Probe pMSU 315 has the advantage of specificity for fungi within the *Gaeumannomyces–Phialophora* group. Although it cannot be used to differentiate between the pathogenic varieties of *G. graminis*, this will often not be a drawback because the two fungi normally occur on different host species. Also, the procedure described does not allow differentiation between *Phialophora* spp.
    GggMR1 can be used to differentiate between the main species and varieties in the group. Weak hybridization of DNA from fungi outside the *Gaeumannomyces–Phialophora* complex sometimes occurs, but the bands for fungi within the group are nevertheless diagnostic (E. Ward, unpublished).

Recent work with GggMR1 has demonstrated the potential of this probe in studies of pathogen population dynamics. Early results suggest that the proportions of the different DNA types of Ggt produced by GggMR1 may vary in a population according to the cereal species grown or the stage in an epidemic.

Southern blotting methods have the drawback of requiring pure cultures of the fungi so that DNA of sufficient quantity and quality can be extracted. DNA extracted from infected roots has been found suitable only in artificially infected plants in which a large proportion of fungal DNA is present (Bateman *et al.*, 1992). Purity is necessary to ensure complete digestion by the restriction enzyme.

Alternative procedures using PCR have been developed recently. These do not require the quantity and high quality of DNA needed by DNA probes and so will allow identification without isolation of pure cultures. A PCR assay specific for *G. graminis* that detects the fungus in infected wheat plants and soil has been described (Schesser *et al.*, 1991; Henson, 1992). This is less effective than probe GggMR1 in discriminating between the different taxonomic groups but, under modified conditions, it can be used to discriminate between the pathogenic and non-pathogenic varieties of *G. graminis* (Ward, 1993). Another PCR-based technique that can be used to discriminate between the various species and varieties in the *Gaeumannomyces–Phialophora* group involves comparison of RFLPs generated by restriction digestion of PCR-amplified ribosomal DNA (Ward, 1993).

REFERENCES

Bateman, G. L., Ward, E. and Antoniw, J. F. (1992) Identification of *Gaeumannomyces graminis* var. *tritici* and *G. graminis* var. *avenae* using a DNA probe and non-molecular methods. *Mycological Research* 96, 737–742.

Henson, J. M. (1989) DNA probe for identification of the take-all fungus, *Gaeumannomyces graminis*. *Applied and Environmental Microbiology* 55, 284–288.

Henson, J. M. (1992) DNA hybridization and polymerase chain reaction (PCR) testing of *Gaeumannomyces, Phialophora* and *Magnaporthe* isolates. *Mycological Research* 96, 629–636.

Holden, M. and Hornby, D. (1981) Methods of producing perithecia of *Gaeumannomyces graminis* and their application to related fungi. *Transactions of the British Mycological Society* 77, 107–118.

Hornby, D., Gutteridge, R. and Parsonage, R. (1991) *Phialophora* spp. in relation to take-all disease of cereals. In: *AFRC Institute of Arable Crops Research Report for 1990*. pp. 73–74.

Juhnke, M. E., Mathre, D. E. and Sands, D. C. (1984) A selective medium for *Gaeumannomyces graminis* var. *tritici*. *Plant Disease* 68, 233–236.

Lee, S. B. and Taylor, J. W. (1990) Isolation of DNA from fungal mycelia and single spores. In: Innis, M. A., Gelfand, D. H., Sninsky, J. J. and White, T. J. (eds) *PCR Protocols: A Guide to Methods and Applications*, Academic Press, San Diego, pp. 282–287.

Marchuk, D., Drumm, M., Saulino, A. and Collins, F. S. (1991) Construction of T-vectors, a rapid and general system for direct cloning of modified PCR products. *Nucleic Acids Research* 19, 1154.

Sambrook, J., Fritsch, E. F. and Maniatis, T. (1989) *Molecular Cloning: A Laboratory Manual*, 2nd edn. Cold Spring Harbor Laboratory Press, New York.

Schesser, K., Luder, A. and Henson, J. M. (1991) Use of polymerase chain reaction to detect the take-all fungus, *Gaeumannomyces graminis*, in infected wheat plants. *Applied and Environmental Microbiology* 57, 553–556.

Walker, J. (1981) Taxonomy of take-all fungi and related genera and species. In: Asher, M. J. C. and Shipton P. J. (eds) *Biology and Control of Take-all.* Academic Press, London. pp. 15–74.

Ward, E. (1993) Use of the polymerase chain reaction for identifying plant pathogens. In: Blakeman, J. P. and Williamson, B. (eds) *Ecology of Plant Pathogens.* CAB International, Wallingford (in press).

Ward, E. and Gray, R. M. (1992) Generation of a ribosomal DNA probe by PCR and its use in identification of fungi within the *Gaeumannomyces–Phialophora* complex. *Plant Pathology* 41, 730–736.

White, T. J., Bruns, T., Lee, S. and Taylor, J. (1990) Amplification and direct sequencing of fungal ribosomal RNA genes for phylogenetics. In: Innis, M. A., Gelfand, D. H., Sninsky, J. J. and White, T. J. (eds) *PCR Protocols: A Guide to Methods and Applications*, Academic Press, San Diego, pp. 315–322.

19 Detection of *Phytophthora fragariae* var. *rubi* in Infected Raspberry Roots by PCR

G. STAMMLER AND E. SEEMÜLLER

Biologische Bundesanstalt für Land- und Forstwirtschaft, Institüt für Pflanzenschutz im Obstbau, Postfach 1214, D-69221 Dossenheim, Germany.

BACKGROUND

Phytophthora root rot is a serious disease of raspberries in North America, Australia and several western European countries. Although a number of *Phytophthora* species have been isolated from diseased roots, one is apparently responsible for the majority of the most damaging outbreaks of root rot worldwide. Morphological criteria, SDS-PAGE protein patterns and molecular genetic studies showed that the pathogen is more closely related to *P. fragariae* than to any other species. Because of some minor differences in morphology and of the host specificity of isolates from raspberry and strawberry, the raspberry pathogen has been formally described as *P. fragariae* var. *rubi* to distinguish it from the agent of red core disease of strawberry *P. fragariae* var. *fragariae* (Wilcox and Duncan, 1993). The disease has been spread worldwide very rapidly via infected plant material. To prevent further spread it is necessary to develop efficient detection procedures. Current methods are based on isolation of the pathogen on selective media or on microscopical methods (Brunner-Keinath and Seemüller, 1992). However, culturing is time consuming and not very sensitive while microscopical detection lacks sensitivity and specifity. Molecular genetics offers a promising alternative for the detection of *P. fragariae* var. *rubi* in infected raspberry roots. Experiments with specific DNA probes used in dot and Southern blot hybridizations were useful for identification and differentiation of *P. fragariae* var. *rubi*, *P. fragariae* var. *fragariae* and other *Phytophthora* species (Stammler *et al.*, 1993). However, it was questionable whether these techniques were sufficiently sensitive for detection. Therefore, the subject of the work was to develop a procedure for the detection of *Phytophthora fragariae* var. *rubi* based on the *in vitro* amplification of fungal DNA by polymerase chain reaction (PCR). Recent studies (Lee and Taylor, 1990; Nazar *et al.*, 1991) showed that small amounts of fungal DNA can be detected by this technique. Because ribosomal DNA is present in multiple copies, it is an attractive target sequence. While the nuclear small-subunit rDNA (ssu or 16S-like) sequences are highly conserved, internal transcribed spacer regions and intergenic spacer are usually poorly conserved and

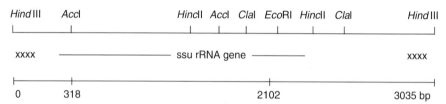

Fig 19.1. Restriction enzyme map of fragment PS3003 containing the small subunit ribosomal RNA gene used for the detection of *P. fragariae* var. *rubi*. xxxx = primer. Primer 1: 5'-CCG TTA CTA GGG GAA TCC TT-3'. Primer 2: 5'-TTC ATT TTC GGA TAG AAC CG-3'.

may vary among species or varieties. These sequence differences offer the opportunity to select species-specific primers for the detection of *P. fragariae* var. *rubi* in host tissue by PCR.

PROTOCOL

Materials

A cloned repetitive DNA fragment of *P. fragariae* var. *rubi* designated PS3003 (Stammler *et al.*, 1993) was used as template DNA. The sequence of fragment PS3003 was compared with corresponding sequences of other oomycetes and non-oomycete fungi available in the EMBL Data Library, Heidelberg, Germany using the HUSAR system. Fragment PS3003 contains the ssu rRNA gene of *P. fragariae* var. *rubi*, which is 1784 bp in size and exhibits a sequence homology of 95.9% with the corresponding gene of *P. megasperma* (Förster *et al.*, 1990). The primers were designed from the flanking regions of the ssu rRNA gene (see Fig. 19.1).
1. DNA isolation buffer (Doyle and Doyle 1990, modified): 2% CTAB, 1.4 m NaCl, 0.2% 2–mercaptoethanol, 20 mm EDTA, 0.1 m Tris-HCl, 1% PVP40 (adjust to pH 7.0 with HCl).
2. $1 \times$ PCR buffer: 100 mM Tris–HCl, 15 mM $MgCl_2$, 500 mM KCl, 0.05% Tween 20, 0.05% NP–40, 0.01% gelatine (adjust to pH 8.3 with HCl).

Method

Step by step protocol: DNA isolation and PCR

1. Wash the roots thoroughly with tap water and blot them with a paper towel.
2. Grind some pieces with a mortar and pestle in 2–3 ml extraction buffer to a homogeneous mash.
3. Put 0.8 ml of the suspension in a 1.5 ml microcentrifuge tube and add the same volume chloroform:isoamylalcohol (24:1).
4. Microcentrifuge at 10,000 g for 5 min at room temperature.
5. Remove the supernatant and dilute it with 0.8 ml H_2O.
6. Apply the solution to a Qiagen tip 20, equilibrated with the recommended equilibration buffer (Qiagen).
7. Wash the Qiagen tip with 2×1 ml recommended washing buffer (Qiagen).
8. Elute the DNA with 0.8 ml of recommended elution buffer (Qiagen).
9. Precipitate the DNA with the same volume of isopropanol and centrifuge it at room temperature for 15 min.

10. Wash the pellet with 80% ethanol, air dry and redissolve it in $100 \mu l$ H_2O.

11. Set up a $50 \mu l$ reaction in a 1.5 ml microcentrifuge tube, mix, and overlay with $50 \mu l$ mineral oil:

$0.5–2.0 \mu l$ DNA preparation (100–400 ng).

$5 \mu l$ 10 × PCR buffer.

$1 \mu M$ primer 1 (see Fig. 19.1).

$1 \mu M$ primer 2 (see Fig. 19.1).

$200 \mu M$ each dNTP.

1.25 units Replitherm (Biozym).

Fill up to $50 \mu l$ with H_2O.

12. Perform 30 cycles of PCR using the following temperature profile:

Denaturation: 94°C, 40 s (first cycle 2 min).

Primer annealing: 50°C, 60 s.

Primer extension: 72°C, 120 s (last cycle 4 min).

13. Analyse the PCR products by separation of $10 \mu l$ of the reaction mixture in 1.0% horizontal agarose gels.

RESULTS

The aim of this work was the development of a PCR method for detecting the raspberry root rot fungus *P. fragariae* var. *rubi* in infected plants. For this purpose a repetitive DNA fragment from *Hind*III-digested genomic DNA of the fungus containing the ssu rRNA gene was selected as template DNA. Primers were selected in terminal regions of this fragment. They enable the amplification of a sequence 3006 bp in size.

With this primer pair, a fragment of the size of the template DNA was obtained when DNA samples from fungal cultures from 13 isolates of *P. fragariae* var. *rubi* and eight isolates of *P. fragariae* var. *fragariae* were amplified (Fig. 19.2). All isolates showed the same restriction pattern when this fragment was digested with each of the following restriction enzymes: *Hinc*II, *Acc*I, *Alu*I, *Cla*I, *Dra*I, *Eco*RI, *Hae*II, *Hinc*II, *Nde*II, *Nru*I, *Rsa*I, *Sau*3A (data not shown). Amplification of DNA from *P. cinnamoni*, *P. syringae*, *P. citricola*, *P. infestans*, *P. cactorum* and *Pythium ultimum* yielded lower amounts of DNA and smaller fragments than that of the two *P. fragariae* varieties. Also, the restriction patterns obtained with the enzymes described above were different from that of the *P. fragariae* varieties and from one another. The detection limit of the system was determined by dilution of DNA from *P. fragariae* var. *rubi*. Amplification was obtained from as little as 50 fg DNA.

The primers were also successfully applied for amplifying the target DNA in samples from infected raspberry roots. A total of 13 naturally infected plants from three different fields, ten plants grown in experimentally infested soil in the field, and eight experimentally inoculated plants from the greenhouse were included in this study. With the exceptions of three plants, a fragment of approximately 3.0 kb in size was detected in all samples regardless of whether the plants showed light or severe symptoms (Fig. 19.3). No DNA was amplified from samples from healthy plants. The identity of the amplified fragment with PS3003 was determined by hybridizing the blotted DNA with a subclone of PS3003, which hybridized with all samples from infected plants. Furthermore, digestion of the amplified DNA with *Acc*I, *Eco*RI and *Hinc*II yielded the same restriction patterns as PS3003 (not shown). No target DNA was amplified from two samples from

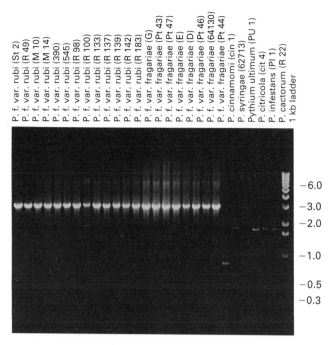

Fig 19.2.   Agarose gel electrophoresis of PCR products from genomic DNA of *P. fragariae* var. *rubi* and *P. fragariae* var. *fragariae* isolates and other *Phytophthora* and *Pythium* species.

non-symptomatic roots of the resistant cultivars. Winklers Sämling and Fallgold grown in experimentally infested soil. Also, one naturally infected plant showing light root rot symptoms tested negative.

LIMITATIONS

The sensitivity of detection by PCR amplification was compared with that of dot and Southern blot hybridization assays. Preliminary experiments showed that of three cloned sequences of *P. fragariae* var. *rubi* (PS39, PS42, PS3003, Stammler., 1992) fragment PS3003 was most sensitive as a probe. However, background reactions made interpretation of dot blot results difficult. On the other hand, fragment PS3003 hybridizes with DNA of *Phytophthora* species other than the two *P. fragariae* varieties (Stammler *et al.,* 1993). For that reason, Southern blot hybridization was used for detection in order to be able to distinguish *P. fragariae* var. *rubi* from other oomycetes by restriction fragment length polymorphism analysis. When the same DNA samples that were used for PCR amplification were subjected to Southern blot hybridization, positive results were obtained with 45% of the plants in contrast to 90% that were detected by PCR (data not shown). *P. fragariae* var. *rubi* is mainly spread with planting and propagation material. As raspberry root rot is difficult to control by fungicide application, the use of healthy plants is the most important measure for disease control. Healthy planting material can only be provided when methods are available to detect light and latent infections. It is likely that the method developed is sufficiently sensitive for this purpose.

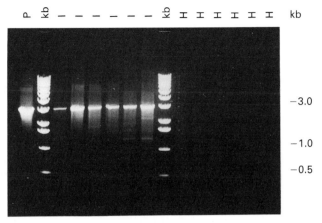

Fig 19.3. Agarose gel electrophoresis of PCR products from DNA from a *P. fragariae* var. *rubi* culture (P), healthy raspberry roots (H), and raspberry roots infected with *P. fragariae* var. *rubi* (I), kb = size marker (1 kb-ladder, Life Technologies).

USE

Because the target sequence is also amplified if DNA from *P. fragariae* var. *fragariae* is used, it should be possible to use the developed PCR method to detect *P. fragariae* var. *fragariae* in infected strawberries (red core root disease) too.

REFERENCES

Brunner-Keinath, S. and Seemüller, E. (1992) Zur Diagnose der *Phytophthora*-Wurzelfäule der Himbeere. *Nachrichtenblatt Deutscher Pflanzenschutzdienst* 44, 179–182.

Doyle, J. J. and Doyle, J. L. (1990) Isolation of plant DNA from fresh tissue. *Focus* (Life Technologies Inc.) 12, 13–15.

Förster, H., Coffey, D. M., Elwood, H. and Sogin, M. L. (1990) Sequence analysis of the small subunit ribosomal RNAs of three zoosporic fungi and implications for fungal evolution. *Mycologia* 82, 306–312.

Lee, S. B. and Taylor, J. W. (1990) Isolation of DNA from fungal mycelia and single spores In: *PCR Protocols, A Guide to Methods and Applications*, Academic Press, San Diego, California.

Nazar, R. N., Hu, X., Schmidt, J., Culham, D. and Robb, J. (1991) Potential use of PCR-amplified ribosomal intergenic sequences in the detection and differentiation of verticillium wilt pathogens. *Physiological and Molecular Plant Pathology* 39, 1–11.

Stammler, G. (1992) Untersuchungen zur molekulargenetischen Differenzierung und zum Nachweis von an Himbeere vorkommenden *Phytophthora*-Arten unter besonderer Berücksichtigung von *Phytophthora fragariae* var. *rubi*. Dissertation, Universität Hohenheim.

Stammler, G., Seemüller, E. and Duncan, J. M. (1993) Analysis of RFLPs in nuclear and mitochondrial DNA and the taxonomy of *Phytophthora fragariae*. *Mycological Research*, 97, 150–156.

Wilcox, W. F. and Duncan, J. M. (1993) *Phytophthora fragariae* var. *rubi varietas* novae. *Mycological Research* (submitted).

20 The Detection and Species Identification of African *Armillaria*

C. MOHAMMED

Department of Plant Sciences, University of Oxford, South Parks Road, Oxford OX1 3RB, UK.

BACKGROUND

Fungal species of the genus *Armillaria* (Fr.:Fr.) Staude can provoke a root rot disease resulting in extensive economic loss to a woody crop. The disease is found worldwide in a range of different hosts and habitats and has been widely reported from the wetter regions of Africa, principally in exotic softwood and cash crop plantations such as coffee, cocoa, rubber, citrus, tea, teak and pine (Hood *et al.*, 1991).

The behaviour of African *Armillaria* was most comprehensively described in the 1960s and 1970s. The literature spans West, Central and East Africa and the attack of various hosts of economic importance. One generality is evident: the stumps of indigenous forest trees harbour latent *Armillaria* lesions which serve as sources of inoculum for the subsequent plantation. It is however easier to detect differences than similarities in the literature of this period. Experience with temperate *Armillaria* suggests that much of this contradictory information might be explained by the presence of several different species. However (as hitherto with temperate *Armillaria*) most cases of *Armillaria* have been arbitrarily attributed to *A. mellea*.

Although much of the research concerning African *Armillaria* is still in progress, significant advances have already been made in the area of taxonomy and sexuality. The methods used to differentiate European *Armillaria* did enable some classification of African *Armillaria* in the preliminary taxonomic studies of Mohammed *et al.* (1989), but they were less discriminative than expected. The pairing test method, so necessary in the determination of the biological species of temperate *Armillaria*, is not entirely transferable to the study of African isolates because many do not have a haploid form. Recent studies include easy and rapid molecular analytical techniques which complement the more 'traditional' investigatory methods, facilitating the task of interpretation.

The publications of Williams *et al.* (1990) and Welsh and McClelland (1990) instigated a great interest in the genetic fingerprinting of organisms by RAPD analysis, a PCR-based technique in which polymorphic sequences are amplified by one random

141

primer. The science is still new but some publications in animal, plant and fungal research indicate its potential for the typing of organisms (Black *et al.*, 1992; Crowhurst *et al.*, 1992; Khush *et al.*, 1992; Wilde *et al.*, 1992). With temperate *Armillaria* species, RAPD analysis has been used to investigate intraspecific differences (Smith *et al.*, 1992). However, the problems faced with unravelling the taxonomy of African *Armillaria* provide an application for RAPDs to be interpreted at an interspecific level.

The banding pattern of many restriction digests, especially digests of genomic DNA, is not visually distinct and RFLP analysis is usually understood as the pattern analysis of probed restriction digests. However, certain restriction enzymes (CG cutters) can cut genomic DNA to give clear banding patterns which may be directly compared on an agarose gel (Croft, Birmingham University, personal communication). This technique of fingerprinting has been applied to the genomic DNA of African *Armillaria*.

Armillaria fungal material is obtained by inoculating a sterile orange segment supported by a glass tube over a column of water with a plug from a culture of African *Armillaria*. The *subcorticalis* type of rhizomorph/mycelial cord growing into the water is harvested 4–6 weeks after inoculation, freeze dried and ground with liquid nitrogen in a pestle and mortar. Over-enthusiastic grinding at this stage can shear DNA. This rhizomorph powder is the material used for all extractions. Several different methods of extracting *Armillaria* genomic DNA were investigated.

PROTOCOLS

Extraction of genomic DNA from African *Armillaria* (technique of Willis *et al.*, 1990)

Materials

1. Extraction buffer (700 mM NaCl, 50 mM Tris-HCl (pH 8.0), 10 mM EDTA, 1% 2-mercaptoethanol (v:v) and 1% CTAB (w:v)).
2. SEVAG (chloroform:isoamylalcohol 24:1).
3. Binding buffer (50 mM Tris, 1 mM EDTA, 6 M NaClO$_4$, pH 7.5).
4. Wash buffer (40 mM Tris, 4 mM EDTA, 0.8 mM NaCl, pH 7.4).
5. Elution buffer (10 mM Tris, 1 mM EDTA, pH 8.0).
6. Prep-A-Gene Matrix (BIORAD).

Method

1. Fill the lower half of the conical portion of a 1.5 ml microcentrifuge tube with powdered mycelium. Add 500 μl of fresh extraction buffer to the powder and mix to produce a homogeneous solution. CTAB is not very soluble in water at room temperature, nor is a 1% solution particularly stable. Make a 10% stock solution and warm it to make the fresh extraction buffer. If the mixture does not flow freely, add more extraction buffer to a maximum of 800 μl.
2. Incubate the mixture at 60°C for 30 min.
3. Emulsify the mixture by adding an equal volume of SEVAG.
4. Centrifuge (12,000 g, 5 min, 25°C, microcentrifuge). Transfer the upper, aqueous phase to a new 1.5 ml microcentrifuge tube. Do not transfer any of the cell debris.
5. Add a quantity of binding buffer to equal three times the combined volume of the

Prep-A-Gene matrix plus DNA sample (10 μl of matrix binds 2 μg of DNA). Mix and leave at room temperature for 5–10 min.

6. Pellet the Prep-A-Gene containing the bound DNA and remove the supernatant. Resuspend and wash the Prep-A-Gene pellet twice with binding buffer.
7. Resuspend and wash the pellet three times with wash buffer.
8. Elute bound DNA with one pellet volume of elution buffer at 45°C.

Extraction of genomic DNA from African *Armillaria* (technique of Möller *et al.*, 1992)

Materials

1. TES (100 mM Tris, pH 8.0, 10 mM EDTA, 2% SDS).
2. Proteinase K, 5 M NaCl.
3. 10% CTAB (cetyltrimethylammomiumbromide).
4. SEVAG (chloroform : isoamylalcohol 24 : 1).
5. 5 M NH$_4$Ac.
6. Isopropanol.

Method

1. Fill the lower half of the conical portion of a 1.5 ml microcentrifuge tube with powdered mycelium. Add 500 μl TES. Add 50–100 μg Proteinase K. Incubate 30–60 min at 55–60°C.
2. Add 140 μl 5 M NaCl and 65 μl CTAB. Incubate for 10 min at 65°C.
3. Add 700 μl SEVAG, incubate at −20°C for 30 min. Centrifuge.
4. Transfer supernantant to another tube. An RNase may be inserted at this point.
5. Add 0.55 volume isopropanol to precipitate DNA. Centrifuge. If no DNA is visible place on ice for about 30 min before centrifugation.
6. Carefully remove supernatant, wash pellet twice with cold 70% ethanol, dry pellet and dissolve it in about 50 μl TE buffer.

RAPDs with genomic DNA from African *Armillaria*

Materials

1. Genomic DNA.
2. Oligonucleotide primers (Oswel DNA Service, Dept. of Chemistry, University of Edinburgh) – primer 1: 5′-ACTTGAGGCG-3′; primer 2: 5′-ATGGATCCGC-3′; primer 3: 5′-CATCATCATCATCAT-3′ (a repeated sequence found in a wide range of organisms (Genbank).
3. Promega *Taq* DNA polymerase, Storage buffer B.
4. dATP, dCTP, dGTP, dTTP (Perkin Elmer).
5. 10 × Reaction buffer; 500 mM KCl, 100 mM Tris-HCl, pH 9.0 at 25°C and 1.0% Triton X-100 (Promega).
6. 25 mM MgCl$_2$.
7. Thermal Minicycler Cycler, Genetic Research Instrumentation Ltd.

Method

1. Perform initial denaturation period of 5 min at 95°C with 23 μl volume including roughly 20 ng genomic DNA, 0.2 μM primer, 100 μM each of dATP, DCTP, dGTP and dTTP, 2.5 μl 10×buffer, 2 mM MgCl₂.
2. Hold at 65°C while adding 0.75 units *Taq* DNA polymerase so that volume of amplication mixture is 25 μl.
3. With primers 1 and 2 follow with 45 cycles of 40°C for 1 min, annealing, 72°C for 2 min polymerization, 94°C for 1 min, denaturation; with primer 3, 35 cycles of 45°C for 1 min, annealing, 72°C for 2 min polymerization, 94°C for 1 min, denaturation.
4. Run amplification 12 μl reaction products on 2% agarose gels, 1×TAE buffer and detect by staining with ethidium bromide.

RFLPs with genomic DNA from African *Armillaria*

Materials

1. Approx 4 μg genomic DNA per sample.
2. Restriction enzymes: *Hae*III, *Msp*I (Sigma).

Method

1. Digest overnight approximately 4 μg of genomic DNA with restriction enzyme.
2. Run digest products on 1% agarose gels (analytical grade) and detect by staining with ethidium bromide.

INTERPRETATION AND LIMITATIONS OF RESULTS

Extraction of genomic DNA

The most common problems encountered in the isolation of pure high MW fungal genomic DNA (50–150 kbp) are the high production by fungi of potential contaminants such as polysaccharides and phenols inhibitory to subsequent analysis. Protocols to efficiently remove such contaminants are few in number. The above two protocols not only have the advantage of effectively eliminating contaminants but are also rapid and give high quality unsheared DNA.

The chromatographic matrix, Prep-A-Gene binds single- and double-stranded DNA in the presence of high concentrations of chaotropes. Since the matrix selectively binds DNA, contaminants including RNA, protein, organic solvents, and other enzyme inhibiting materials are removed by the brief washing step.

The Möller *et al.* (1992) technique combines the inactivation of proteins by SDS/proteinase K with precipitation of acidic polysaccharides by hot CTAB in the presence of SDS and high salts and only a single selective precipitation of DNA with isopropanol.

Limitations

The technique of Willis *et al.* (1990) is only used to extract small quantities of DNA for RAPD analysis due to the high cost of the matrix.

RAPDS with genomic DNA from African *Armillaria*

As in the North American *Armillaria* studies of Smith *et al.* (1992) primers 1 and 2 give multiple fragments segregating for presence or absence in the diploid isolates of European and African *Armillaria*. Primer 3 produces one or two segregating fragments. Intraspecific differences between French, Scottish and Polish control isolates of a well-defined European *Armillaria* species do not mask a distinct interspecific profile (Fig. 20.1). African *Armillaria* isolates can be grouped for each primer according to pattern similarities (Fig. 20.1). The isolates within these different groups are of widely different geographic origin and it is probable that these isolates do constitute a biological group or species. The degree to which intraspecific variation is expressed in a particular group of isolates varies with each primer. However, all three primers give corresponding groupings. Different types of taxonomic investigation and species characterization (observation of vegetative mat morphology in culture, production of subterranean rhizomorphs, fruiting in culture, ability to fruit, morphology of fruitbodies, study of monospore isolates, mating tests) integrate well with the definition of biological groups based on RAPD analyses. In addition, newly collected isolates, not assigned to groups by traditional means, were distinguished by RAPD patterns.

Limitations

Many factors influence RAPD patterns: DNA quality, purity and concentration, magnesium concentration, polymerase concentration and denaturing temperature. The overriding and most important source of variability in these studies was any deterioration in DNA quality and purity.

The successful application of protocols optimized in other laboratories is difficult. Once a working protocol has been developed in a particular laboratory it is advisable to use two controls for each reaction; amplification mixture without DNA (to check for contamination) and amplification mixture with a control DNA (a change in pattern with any one primer will indicate variability in amplification conditions).

The interpretation of patterns at an interspecific level is open to subjective interpretation. A cluster analysis of isolates based on band present/band absent scores is envisaged.

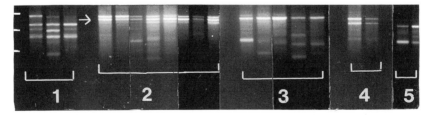

Fig 20.1. RAPD analysis of genomic DNA from isolates of *Armillaria* from Europe and West, Central and East Africa: Primer 5′-ACTTGAGGCG-3′. Lanes from left to right: position of base pair markers: 1353, 603, 234; group 1: French, Scottish and Polish isolates of *A. ostoyae*; group 2: three isolates from Tanzania, three isolates from La Réunion, three isolates from Kenya; group 3: isolates from the Ivory Coast and Liberia, three isolates from Gabon; group 4: isolates from Zimbabwe; groups 2, 3 and 4 all share a polymorphic fragment (marked with arrow) and are considered as possible subgroups of a species *sensu lato*; group 5: isolates from above 2000 m in Kenya.

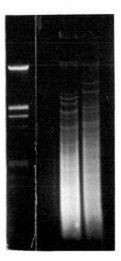

Fig 20.2.   Genomic DNA of African *Armillaria* isolates cut with *Hae*III. Lanes from left to right: base pair markers: 21,226, 5184, 4973, 4277, 3530, 2027, 1904; Tanzanian isolate from group 2; Kenyan isolate from group 5.

RFLPs with genomic DNA from African *Armillaria*

Restriction enzymes *Msp*I and especially *Hae*III cut African *Armillaria* genomic DNA to give clear banding patterns which can be directly compared on an agarose gel (Fig. 20.2). This technique of fingerprinting can be applied to the genomic DNA of African *Armillaria* in tandem with RAPDs.

Limitations

This technique requires a far larger quantity of DNA than RAPDs although this may easily be obtained using Möller *et al.*'s (1992) genomic extraction method.

REFERENCES

Black, W. C. IV, Duteau, N. M., Puterka, G. K. Nechols, J. R. and Pettorini, J. M. (1992) Use of random amplified polymorphic DNA polymerase chain reaction (RAPD-PCR) to detect DNA polymorphisms in aphids (Homoptera: Aphididae). *Bulletin of Entomological Research* 82, 151–159.

Crowhurst, R. N., Hawthorne, B. T., Rikkerink, E. H. A. and Templeton, M. D. (1992) Differentiation of *Fusarium solani* f.sp. *curcurbitae* races 1 and 2 by random amplification of polymorphic DNA. *Current Genetics* 20, 391–396.

Hood, I. A., Redfern, D. B. and Kile, G. A. (1991) *Armillaria* in planted hosts. In: Shaw, C. G. and Kile, G. A. (eds) *Armillaria Root Disease*. United States Department of Agriculture Forest Service Agriculture handbook no. 691. Forest Service, USDA, Washington, DC, pp. 122–149.

Khush, R. S., Becker, E. and Wach, M. (1992) DNA amplification polymorphisms of the cultivated mushroom *Agaricus bisporus*. *Applied and Environmental Microbiology* 58, 2971–2977.

Mohammed, C., Guillaumin, J. J. and Berthelay, S. (1989) Preliminary investigation about the taxonomy and genetics of African *Armillaria* species. In: Morrison, D. J. (ed.) *Proceedings of the*

Seventh International Conference on Root and Butt Rots. IUFRO Working Party S2.06.01, British Columbia, August 1988. Forestry Canada, Pacific Forestry Centre, Victoria, BC, pp. 447–457.

Möller, E. M., Bahnweg, G., Sandermann, H. and Geiger, H. H. (1992) A simple and efficient protocol for isolation of high molecular weight DNA from filamentous fungi, fruit bodies and infected plant tissues. *Nucleic Acids Research* 20, 6115–6116.

Smith, M. L., Bruhn, J. N. and Anderson, J. B. (1992) The fungus *Armillaria bulbosa* is among the largest and oldest living organisms. *Nature* 356, 428–431.

Welsh, J. and McClelland, M. (1990) Fingerprinting genomes using PCR with arbitrary primers. *Nucleic Acids Research* 18, 7213–7218.

Wilde, J., Waugh, R. and Powell, W. (1992) Genetic fingerprinting of *Theobroma* clones using randomly amplified DNA markers. *Theoretical and Applied Genetics*, 83, 871–877.

Williams, J. G. K., Kubelik, A. R., Livak, K. J., Rafalski, J. A. and Tingey, S. V. (1990) DNA polymorphisms amplified by arbitrary primers are useful as genetic markers. *Nucleic Acids Research* 18, 6531–6535.

Willis, E. H., Mardis, E. R., Jones, W. L. and Little, M. C. (1990) Prep-A-Gene™. A superior matrix for the purification of DNA and DNA fragments. *Biotechniques* 9, 92–99.

21 The Development of Monoclonal Antibody-based ELISA and Dipstick Assays for the Detection and Identification of *Armillaria* Species in Infected Wood

R. Priestley,[1] C. Mohammed[2] and F. M. Dewey[2]

[1] *Present address, Delta Biotechnology, Castle Court, 59 Castle Boulevard, Nottingham NG7 1FD, UK:* [2] *Department of Plant Sciences, University of Oxford, South Parks Road, Oxford OX1 3RB, UK.*

Background

For over a century, the confusion surrounding the nomenclature and taxonomy of the genus *Armillaria* greatly hindered the study of this fungal pathogen. Until the 1980s most pathologists attributed the disease syndrome to a single, highly variable *Armillaria* species, *A. mellea* (Vahl: Fr.) Kummer. However, recent research in Europe, North America and Australia has developed and exploited pairing tests based on somatic incompatibility (unknown diploid or haploid isolates are paired with identified haploid tester strains) to establish a considerable number of 'biological species' with different distributions, host ranges, pathogenicity and cultural characteristics (Guillaumin *et al.*, 1989, 1991).

In Europe *A. mellea* (Vahl:Fr.) and *A. ostoyae* (Romagnesi) Herink are both highly pathogenic, but *A. mellea* is predominantly found in deciduous forest, orchards and vineyards and *A. ostoyae* in conifers. Other species (*A. gallica* Marxmüller, *A. cepistipes* Velenovsky, *A. borealis* Marxmüller and Korhonen, *A. tabescens* (Scop.:Fr.) Emel.) are saprophytes or weak parasites.

Successful isolation of *Armillaria* in pure culture is a prerequisite to the pairing test method of identification. Although in such common use, pairing tests are long and tedious and sometimes open to subjective interpretation. The development of an immunological user-friendly identification kit, both objective and rapid, would greatly facilitate *Armillaria* research, especially if *Armillaria* could be detected and identified from infected wood. Such an advance would not only be of importance to the scientist but of practical benefit to the forester in influencing sylvicultural management. Studies have therefore focused on the immunological detection and identification of the two

149

economically important pathogenic species of European *Armillaria*: *A. ostoyae* and *A. mellea*.

Previous identifications of different *Armillaria* species with polyclonal antisera in ELISA (Lung-Escarmant *et al.*, 1985) and with monoclonal antibodies derived from a 'shotgun' method of inoculating a mouse with crude mycelial extracts (Fox and Hahne, 1989) showed quantitative differences between species. The value of such methods was evident but they needed to be extended and refined to give qualitative differences by obtaining species-specific monoclonal antibodies. Only if the specificity of identification could be augmented would the development of a user-friendly assay for general diagnostic use by untrained personnel in the field or in the laboratory become more than an ideal possibility.

The techniques used at the Department of Plant Sciences, University of Oxford in raising species- and genus-specific MAbs to fungal pathogens such as *Ophiostoma ulmi*, *Humicola lanuginosa*, *Penicillium islandicum* and *Botrytis cinerea* (Dewey *et al.*, 1989; Bossi and Dewey, 1992; Dewey, 1992) were applied to developing MAbs which would recognize different species of European *Armillaria*.

Antibodies were raised to *A. mellea* and *A. ostoyae* by either of two ways:

1. Immunizing mice with low molecular weight fractions excised from SDS polyacrylamide gels. Proteins extracted from lyophilized rhizomorph powder were separated by SDS polyacrylamide gel electrophoresis (SDS-PAGE). Most of the differences between the species were found to occur within the 21–14 kD region and it was this low molecular weight region of the gel that was used as the antigen to immunize mice.

2. Immunizing mice with proteins precipitated from extracts from lyophilized rhizomorph powder. Proteins were extracted from lyophilized mycelial powder (see *Armillaria* RAPDs section on fungal culture) with 30 mM Tris-HCl (pH 7.2) containing 20% glycerol and 1 mM β-mercaptoethanol. After elimination of debris, the supernatant was precipitated with 5 volumes of cold acetone. The precipitate was then dried and resuspended in phosphate buffered saline (PBS). Mice were immunized with 30 μg of protein per injection.

Monoclonal antibodies arising from the fusions carried out with mice injected with either gel proteins or precipitated proteins were screened by ELISA against PBS extracts from lyophilized rhizomorph powder. A fusion with a mouse immunized with *A. ostoyae* gel proteins (21–14 kD) gave rise to 11 monoclonal antibodies that recognized *A. ostoyae*, two of which (FG6 and GG12) were found to be species-specific (Table 21.1) when tested against other *Armillaria* species and other wood-inhabiting fungi. A fusion with a mouse immunized with precipitated proteins also gave rise to an *A. ostoyae* specific antibody (JE12) (Table 21.1).

Fusions with mice immunized with *A. mellea* gel proteins (21–14 kD) gave rise to seven monoclonal antibodies that recognize *A. mellea*. Two of these (ID12 and AA3) were found to be species-specific, and two others (BA4 and AH1) were found to be genus-specific (Table 21.1).

Immunoblotting carried out using the *A. ostoyae*-specific antibody (FG6) (Fig. 21.1), shows that it binds strongly to bands within the 21–14 kD region used to immunize the mice and in particular to an *A. ostoyae* band within this region.

The antibodies that were species-specific and genus-specific, when screened against

Table 21.1. Monoclonal antibodies raised to *A. ostoyae* and *A. mellea* gel proteins and precipitated proteins. Screened against PBS extracts from rhizomorph powder.

	BA4	AH1	ID12	AA3	FG6	JE12	GG12
A. mellea							
D2	+	+			−		
D12	+	+	+	+	−	−	−
D16	+	+			−		
A. ostoyae							
C1	+	+	−	−	+	+	+
C6	+	+			+		
C9	+	+			+		
A. borealis	+	+	ND	ND	−	ND	ND
A. gallica	+	+	−	−	−	−	−
A. cepistipes	+	+	−	−	ND	−	−
A. tabescens	+	+	−	−	ND	−	−
Phaeolus schweinitzii	−	−	ND	ND	−	ND	ND
Ophiostoma ulmi	−	−	ND	ND	−	ND	ND

ND = not determined.

PBS extracts from rhizomorph powders, were then screened against extracts from infected wood material.

PROTOCOLS

ELISA assay for detecting and identifying *Armillaria* in wood

Materials

1. Multiwell polystyrene strips and strip holder frames (Microstrips and Combi frames, Labsystems UK).
2. Antigen extraction buffer – phosphate buffered saline (PBS) pH 7.2.
3. Strong, sharp scalpel or razor blade.
4. Wash buffer (PBST) – PBS with 0.05% Tween 20.
5. Goat anti-mouse IgG+IgM peroxidase conjugate (Sigma 0412).
6. Substrate buffer (5 ml 0.2 M sodium acetate, 195 μl 0.2 M citric acid made up to 10 ml with H_2O)+5 μl H_2O_2 and 100 μl TMB stock substrate solution (10 mg ml^{-1} 3,3′,5,5′-tetramethyl-benzidine in dimethyl sulphoxide, stored at 4°C).
7. Test material (wood with obvious fungal infection, suspected or known to be *Armillaria* infection).

Method

1. Take shavings from the test material using a strong, sharp scalpel. Take as much of the visible fungal material as possible.
2. Incubate shavings in extraction buffer (1/3 w/v) for 6 h with rolling or shaking.

Table 21.2. Results of detection and identification by MAb ELISA, pairing tests and observation of culture morphology.

Fungal material/mycelium	MAb test result	Morphology in pure culture	Pairing test result
Oxfordshire BW1 – in rotting stump	+ve CA12, −ve JE12	*mellea*	*mellea*
Oxfordshire BW2 – in rotting stump	?	*gallica*	*gallica*
Oxfordshire BW3 – in rotting stump	+ve JE12, −ve FG6 and −ve CA12	*gallica*	*gallica*
Oxfordshire Wh1 – in rotting stump	+ve CA12, −ve JE12	Isolation unsuccessful	
Oxfordshire Wh2 Rhizomorphs on stump	+ve CA12, −ve JE12	Isolation unsuccessful	
Oxfordshire Hinksey 1 – in dead willow	+ve CA12, −ve JE12	*mellea*	*mellea*
Suffolk, Studhall – in dying oak	+ve JE12, −ve CA12	*gallica*	*gallica*
Lincolnshire, Grimsthorpe in dead oak	+ve JE12,−ve CA12	*mellea/gallica*	*mellea*
Yorkshire, Huddersfield – in dead elm, rotted material and rhizomorphs	+ve CA12, −ve JE12	Isolation unsuccessful	
Yorkshire, Huddersfield – in dead ash	+ve CA12, −ve JE12	Isolation unsuccessful	
Gloucestershire, Forest of Dean – in Sitka spruce	+ve JE12, −ve CA12	*ostoyae*	*ostoyae*
Surrey, Farnham – in dying Leyland cypress	+ve CA12,−ve JE12	*mellea*	*mellea*
Scotland, Glentress – in Sitka spruce	+ve JE12,−ve CA12	*ostoyae/borealis*	*ostoyae*
Scotland, Bush – in Sitka spruce	−ve JE12,−ve CA12	*gallica/cepistipes*	*gallica*
Auvergne, France – known *ostoyae* No 148, Clone 7	+ve JE12, −ve CA12	—	Previously identified
Auvergne, France – known *ostoyae* No 151, Clone 8	+ve JE12, −ve CA12	—	Previously identified
Bordeaux, France, known *ostoyae*	+ve JE12,−ve CA12, +ve FG6	—	Previously identified
Didcot, rotted wood	−ve JE12, −ve CA12	Isolation unsuccessful	
Lime, rotted wood	−ve JE12, −ve CA12	Isolation unsuccessful	

CA12 is *A. mellea* specific; JE12 recognizes *A. gallica* and *A. ostoyae*;
FG6 is *A. ostoyae* specific.

3. Spin down the shavings for 5 min in a bench centrifuge. The supernatant may have to be spun in a microcentrifuge to remove all the debris.
4. Coat microlitre wells with the extract (supernatant) with double dilutions, in to PBS across the plate ($50\,\mu$l per well).
5. After incubating the plate overnight at 4°C, invert and flick out residual antigen solution.
6. Wash four times with the washing buffer, once with PBS and briefly with water. Dry the plate in a laminar flow hood and store sealed at 4°C until ready for processing.

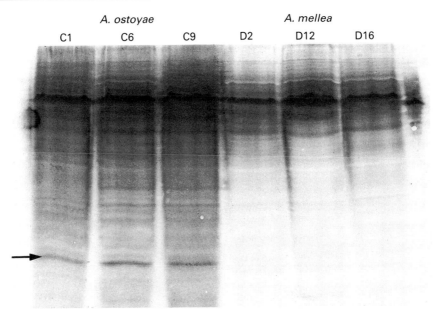

Fig 21.1.   Western blot from an SDS polyacrylamide gel of proteins precipitated from mycelial powder of three isolates of *A. ostoyae* (C1, C6, C9) and three isolates of *A. mellea* (D2, D12, D16). The blot was probed with the *A. ostoyae* monoclonal antibody FG6 and binding visualized using an antimouse gold conjugate. An *A. ostoyae*-specific band is indicated by an arrow.

7. Block the plate by incubating with 1% BSA for 10 min (100 μl per well).
8. Wash plate four times with PBST.
9. Incubate separate antigen-coated strips with the *A. mellea*-specific monoclonal antibody, the *A. gallica* and *A. ostoyae* antibody and the *A. ostoyae*-specific antibody (50 μl per well, for 1 h).
10. Wash plate four times with PBST.
11. Incubate with goat antimouse peroxidase conjugate (Sigma 0412) diluted 1 in 1000 into PBST (50 μl per well for 1 h).
12. Wash four times with PBST.
13. Add 50 μl of substrate solution to each well, incubate 15 min.
14. Stop reaction by addition of 50 μl of 5 M H_2SO_4 to each well. Read absorbency at 450 nm on an ELISA plate reader.

Dipstick assay for detecting and identifying *Armillaria* in wood

Materials

1. Strips, 5 × 40 mm of Millipore (PVDF) membrane (Immobilon P).
2. 100% methanol
3. Blocking/washing/diluent buffer: Tris-HCl pH 7.6, 0.3% casein
4. Goat anti-mouse gold conjugate (Biocell)
5. Silver enhancer kit (Biocell)

Method

1. Incubate dipsticks overnight in fungal extraction supernatant diluted 1/192 (step 3 ELISA assay).
2. Air dry strips for 15 min.
3. Code strips (with nicks). The strips are then processed together in a single tube.
4. Re-wet in 100% methanol for 2 min, rinse with distilled water for 1 min.
5. Agitate at room temperature in blocking solution for 1 h.
6. Expose to primary antibody supernatant (*A. mellea-* or *A. ostoyae*-specific)+0.1% casein.
7. Wash with three 5 min rinses.
8. Incubate with goat antimouse gold conjugate diluted 1/100 for 1 h.
9. Visualize using silver enhancer (20 min).

Interpretation and Limitations of Results

In preliminary assays of the antibodies screened against infected wood material, antibody CA12 proved to be specific for *A. mellea*, JE12 recognizes both *A. gallica* and *A. ostoyae* and FG6 is specific for *A. ostoyae*.

Antibodies CA12 and JE12 were then screened against a wider range of infected material; this included wood clearly infected with *Armillaria* and an unidentified wood rot suspected to be of fungal origin. Isolations on to a selective culture medium were attempted with all samples (Table 21.2). The presence of *Armillaria* was detected immunologically even in a sample of rotted wood with no visually evident mycelium and from which the fungus could not be successfully isolated.

The *Armillaria* cultures isolated were identified by their morphology in culture and pairing tests. Identifications with MAbs and pairing tests are compared in Table 21.2.

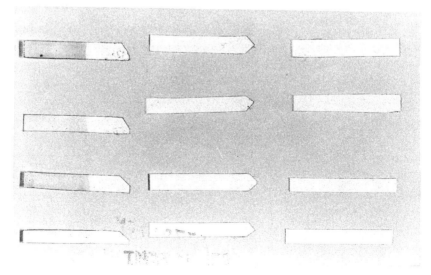

Fig 21.2. Dipsticks incubated in an extract from *A. mellea*-infected wood.

There is only one case of discrepancy between the results given with MAbs and pairing tests.

Dipstick assays with MAbs CA12 and JE12 detected and identified *A. mellea* and *A. gallica* in infected wood (Fig. 21.2). These results were corroborated by pairing tests.

The comparison in the steps and time taken for detection and identification of *Armillaria* by pairing tests and ELISA or dipstick assay are compared below and show the obvious advantages and ease of the immunological assay.

Immunological-based assay	Pairing test
Shavings from infected wood – 20 min	Preparation of selective medium – 3 h
Fungal antigens extracted in PBS – 6 h	Isolation – 10 min per sample
Coat ELISA wells or incubate dipsticks with antigen supernatant – overnight	Growth of culture on selective medium– 10–15 days Transfer to and growth on malt/agar – 10–15 days
Probe with MAbs – 1 h	Preparation for pairing tests of 18 Petri dishes + malt/agar for each sample – 3 h
Detect MAbs with peroxidase conjugate (ELISA)	Pairing tests with haploid testers or gold conjugate (dipsticks) – 1 h 15 min per sample
Visualize using TMB (ELISA) or silver enhancer (dipsticks) – 20 min	Pairing tests ready to interpret after 15–20 days
Total: under 24 h	Total: 6–8 weeks

Limitations

It is necessary to test against dilution series of the antigen to ensure specificity at species level. Assays with extractions from infected wood which has been allowed to dry are not successful. However, if the dried material is re-wetted and the *Armillaria* resumes active growth then antigenic extractions may be obtained.

Care must be taken to avoid contact with the membranes in the dipstick assay. Silver enhancement can be variable in its results and this stage must be constantly observed.

Although the MAb-based ELISAs of *Armillaria*-infected wood show great potential, more vigorous testing of a wide range of wood-rotted material by *Armillaria* and other fungi must be screened with ELISA before it may be possible to exploit the dipstick assay commercially.

References

Bossi, R. and Dewey, F. M. (1992) Development of a monoclonal antibody based immunodetection assay for *Botrytis cinerea*. *Plant Pathology* 41, 472–482.

Dewey, F. M. (1992) Detection of plant invading fungi by monoclonal antibodies. In: Duncan, J. M. and Torrance, L. (eds) *Techniques for the Rapid Detection of Plant Pathogens*. Blackwell Scientific, Oxford, pp. 47–62.

Dewey, F. M., MacDonald, M. M. and Phillips, S. I. (1989) Development of monoclonal-antibody-ELISA, DOT-BLOT and DIP-STICK immunoassays for *Humicola lanuginosa* in rice. *Journal of General Microbiology* 135, 361–374.

Fox, R. T. V. and Hahne, K. (1989) Prospects for the rapid diagnosis of *Armillaria* by monoclonal antibody ELISA. In: Morrison, D. J. (ed.) *Proceedings of the Seventh International Conference on Root and Butt Rots, Vernon and Victoria, British Columbia*, 1988 IUFRO Working Party S2.06.01. Pacific Forestry Centre, Victoria, British Columbia, pp. 458–469.

Guillaumin, J. J., Mohammed, C. and Berthelay, S. (1989) *Armillaria* species in the Northern temperate hemisphere. In: Morrison, D. J. (ed.) *Proceedings of the Seventh International Conference on Root and Butt Rots*. IUFRO Working Party S2.06.01, British Columbia, August 1988, Pacific Forestry Centre: Victoria, BC, pp. 27–44.

Guillaumin, J. J., Anderson, J. B. and Korhonen, K. (1991) Life cycle, interfertility, and biological species. In: Shaw, C. G. and Kile, G. A. (eds) *Armillaria Root Disease*. United States Department of Agriculture Forest Service Agriculture handbook no. 691. Forest Service, USDA, Washington, DC, pp. 10–20.

Lung-Escarmant, B., Mohammed, C. and Dunez, J. (1985) . Nouvelles méthodes de détermination des Armillairés européens: Immunologie et electrophorèse en gel de polyacrylamide. *European Journal of Forest Pathology* 15, 278–288.

22 Development of an ELISA and Dot-blot Assay to Detect *Mucor racemosus* and Related Species of the Order Mucorales

G. A. De Ruiter,[1] W. Bos,[2] A. W. Van Bruggen-Van der Lugt,[1] H. Hofstra[2] and F. M. Rombouts[1]

[1] *Department of Food Science, Wageningen Agricultural University, PO Box 8129, 6700 EV Wageningen, The Netherlands;* [2] *Institute for Biotechnology and Chemistry, TNO-Food Division, PO Box 360, 3700 AJ Zeist, The Netherlands.*

BACKGROUND

Moulds belonging to the order Mucorales (class Zygomycetes) are relatively primitive fungi characterized by coenocytic mycelium and by the production of solitary sexual spores, called zygospores. Many species of Mucorales are widely distributed on food products, particularly on stored grain, fruits and vegetables and cause many kinds of food spoilage (Pitt and Hocking, 1985; Samson and Van Reenen-Hoekstra, 1988). *Rhizopus stolonifer* (Ehrenb.) Lind, often nicknamed as 'bread mould', is the most commonly occurring species and is frequently the cause of postharvest deterioration of fruits, vegetables and cereals. This species is mildly pathogenic to maturing fruit before harvesting, but primarily a wound-invading mould. The genus *Mucor* (e.g. *M. racemosus* Fres., *M. hiemalis* Wehmer and *M. circinelloides* van Tieghem) has a worldwide distribution and species are isolated very often from a broad range of foods. The equally cosmopolitan *Absidia corymbifera* (Cohn) Sacc. and Trotter is isolated frequently from cereals, cereal products and decaying fruits and vegetables. Due to the high optimum temperature of growth (approx. 38°C) this species commonly occurs on heated substrates and can be pathogenic to man and cattle. *Syncephalastrum racemosum* Cohn, found mainly in tropical regions, has also a relatively high optimum temperature of growth. *Thamnidium elegans* Link is a psychrophilic mould commonly associated with cold-stored meat. Some species of Mucorales are important for use in fermented foods, e.g. *Rhizopus oligosporus* Saito in tempe production. Some species are important as causal agents of mucormycosis in man (Sugar, 1992). Furthermore, some species of *Choanephora* (e.g. *C. cucurbitarum* (Berk. and Rav.) Thaxter) are pathogenic for crop plants.

In general, the presence of moulds in food is often associated with loss of quality, therefore their early detection is important. The methods commonly used for detection of moulds in food products have many disadvantages (Jarvis *et al.*, 1983). The microscopic methods for detecting mycelium, such as the Howard Mould Count (HMC), lack precision. The chemical methods are particularly laborious and also lack precision. Plating techniques (mould colony count), based on the enumeration of viable propagules including asexual and/or sexual spores, are not very reliable either. Mycelium, even when present in large amounts, usually leads to low colony counts.

Modern detection methods in food mycology are based on the recognition of fungal antigenic polysaccharides by specific immunoassays (Notermans and Heuvelman, 1985; De Ruiter *et al.*, 1993a). Recently, the development of a specific ELISA for species of Mucorales was described and successfully used to detect these species in food (De Ruiter *et al.*, 1992a). This sandwich ELISA is based on polyclonal IgG antibodies (PAbs) raised in rabbits against the extracellular polysaccharides (EPS) of *Mucor racemosus* which are almost specific for moulds belonging to the order of Mucorales, as cross-reactivity was only observed with the yeast *Pichia membranaefaciens* (De Ruiter *et al.*, 1992b). To enhance the specificity and to allow the commercialization of specific test kits, monoclonal antibodies (MAbs) are necessary (Dewey, 1992). Recently, MAbs have been described specific for *Penicillium, Aspergillus* and *Botrytis* species and used in the analysis of foods and feeds (Dewey *et al.*, 1990; Bossi and Dewey, 1992; Stynen *et al.*, 1992).

In the present chapter, we describe the development of an indirect ELISA and a dot-blot assay based on a monoclonal IgG antibody against EPS of *Mucor racemosus* (De Ruiter *et al.*, 1993b). These assays were used to determine the specificity of the MAb and to compare its activity with that of polyclonal IgG antibodies raised in rabbits against the same EPS.

PROTOCOLS

Preparation of the antigens

The strain of *M. racemosus* was grown on a yeast nitrogen base (YNB, Difco Labs, Detroit, USA) synthetic culture medium supplemented with D-glucose (Merck) as carbon source in a concentration of $30 \, \text{gl}^{-1}$. After incubation for 7 days at 30°C on a rotary shaker at 100 rpm, the culture liquid (200 ml) was separated from the mycelium by filtration. The filtrate was concentrated fivefold with a vacuum rotary evaporator and heated for 5 min at 100°C to inactivate enzymes which might be present. The filtrate was poured into a dialysis bag, dialysed against running tap water overnight, then against distilled water for 24 h and lyophilized. The residue was dissolved in 20 ml of distilled water, any water-insoluble material was removed by centrifugation and the EPS-containing water fraction was poured into five volumes of ethanol (96%) and stored for 16 h at 4°C. Finally, the precipitate was separated by centrifugation (30 min, 19,600 g), dissolved in 20 ml of distilled water, reprecipitated with ethanol once and lyophilized, giving the water-soluble antigenic EPS (De Ruiter *et al.*, 1991). Polyclonal (pAb 1000/1201) and monoclonal IgG antibodies (MAb 12.8) raised against these antigens were prepared as described recently by De Ruiter and co-workers (1993b).

Indirect ELISAs

Materials

1. Polyvinyl 96-well microtitre plates (Dynatech, Chantilly, VA, USA).
2. Washing buffer PBST: PBS with 0.05% (w/v) Tween 20.
3. Microplate washer: Titertek S8/12 (Flow Laboratories).
4. (a) Peroxidase-conjugated goat antimouse antibodies (Sigma Immunochemicals A-4416; St. Louis, MO, USA).
   (b) Peroxidase-conjugated goat antirabbit antibodies (Sigma A-8275).
5. Substrate solution prepared just before use: 10 ml of a 0.1 M sodium acetate/citric acid buffer (pH 6.0), 100 μl of 42 mM TMB (10 mg ml^{-1} 3,3′,5,5′-tetramethyl benzidine in DMSO), 0.5 μl H$_2$O$_2$ (Bos *et al.*, 1981).
6. 2 M sulphuric acid.
7. ELISA reader: EAR 400 spectrophotometer (SLT, Groedig, Austria) at λ450 nm.

Method

1. 10 g of a sample are diluted with 90 ml of PBS or peptone physiological saline (PPS; 1 g l^{-1}, pH 7.2, 150 mM NaCl) and homogenized with a stomacher mixer (Seward 400, London, UK) for 1 min.
2. (a) Wells of the microtitre plates are coated with 100 μl of the supernatant after centrifugation (15 min, 3000 rpm) of the extracts, and incubated for 16 h at 25°C.
   (b) One row on a microtitre plate is coated with 100 μl of a 5 μg ml^{-1} preparation of EPS from *Mucor racemosus* in PBS for calibration.
3. The plates are washed three times with PBST.
4. 100 μl of the respective antibody (approx. 10 μg ml^{-1}) in PBS containing 1% (w/v) BSA are added to each well and incubated for 1.5 h at 25°C.
5. The plates are washed three times with PBST.
6. The respective peroxidase-conjugate is diluted 1000-fold in PBST containing 1% (w/v) BSA and 100 μl are subsequently added to each well and incubated for 1.5 h at 25°C.
7. The plates are washed three times with PBST.
8. 100 μl of TMB substrate solution are added to each well and incubated for 30 min at 25°C.
9. 50 μl of 2 M H$_2$SO$_4$ are added to each well.
10. The absorbance of the yellow colour is measured spectrophotometrically.

Dot-blot assay

Materials

1. Nitrocellulose paper 0.1 μm (Schleicher and Schüll, Dassel, Germany).
2. Blocking agent: sodium caseinate (DMV Campina, Veghel, The Netherlands) dissolved in PBS.
3. (a) Peroxidase-conjugated goat antimouse antibodies (Sigma A-4416).
   (b) Peroxidase-conjugated goat antirabbit antibodies (Sigma A-8275).
4. Substrate solution: PrestoSol BL, ready-to-use 4-chloro-1-naphthol containing solution (Janssen Biotech NV, Olen, Belgium).

5. Optical densitometer: Computing densitometer 300A (Molecular Dynamics, Sunny-vale, CA, USA) equipped with a helium-neon laser (λ 672 nm) using ImageQuant 3.22 software.

Method

1. (a) Test material extracted in PBS is bound to the nitrocellulose membranes by spotting droplets of $2\,\mu l$ to the membrane.
   (b) A solution of purified EPS ($5\,mg\,ml^{-1}$) is used as a positive control.
2. After the test material has been allowed to air-dry for 30 min at 25°C it is blocked with a 0.1% (w/v) solution of caseinate for 1 h at 25°C.
3. The membrane is exposed to a solution of $2.5\,\mu g\,ml^{-1}$ MAb IgG in PBS containing 0.5% BSA for 16 h at 25°C.
4. The membrane is washed with PBST, three times, 2 min.
5. The membrane is exposed to the respective peroxidase-conjugate, diluted 1000-fold in PBST containing 1% (w/v) BSA for 1.5 h at 25°C.
6. The membrane is washed with PBST, three times, 2 min.
7. The membrane is put in the ready-to-use 4-chloro-1-naphthol solution for 30 min at 25°C.
8. The optical density of the blue spots is measured with the densitometer.

RESULTS

The specificity of MAb 12.8 was studied by determination of the titres of 29 different EPS preparations of the mould genera *Mucor, Rhizopus, Rhizomucor, Absidia, Syncepha-lastrum, Thamnidium, Mortierella, Penicillium, Aspergillus, Fusarium, Alternaria, Botrytis* and *Cladosporium*. As can be derived from Table 22.1, MAb 12.8 gave a positive immunoreac-tion with the EPS preparations from all moulds belonging to the order Mucorales tested, except species of the genus *Mortierella sensu stricto* which includes the species *reticulata, hyalina* and *polycephala* (De Ruiter *et al.*, 1993c). No reactivity was observed with other mould species often occurring in food such as *Penicillium, Aspergillus* and *Fusarium*. Furthermore, 39 yeast species belonging to 20 different genera (see Table 22.1) were tested for cross-reactivity with MAb 12.8. These yeasts did not show any reactivity with MAb 12.8, including the yeast *Pichia membranaefaciens*, which was the sole genus to give a clear positive immunoreaction with PAb 1000/1201 (Table 22.1).

The reactivity of MAb 12.8 was compared with the reactivity of PAb 1000/1201 antibodies raised in rabbits against the same EPS preparation of *M. racemosus* as described previously (De Ruiter *et al.*, 1992b). Both antibodies reacted with all EPS obtained from species of the order Mucorales tested, but their reactivity differed significantly. PAb 1000/1201 gave the highest reactivity with the EPS preparation of *Mucor circinelloides* whereas MAb 12.8 gave only a weak immunoreaction with this EPS. In contrast, the EPS from *Rhizomucor* species showed a much higher reaction with MAb 12.8 than with PAb 1000/1201. To exclude the possibility that these differences in reactivity were caused by differences in binding properties of the EPS preparations to the wall of the microtitre plates, a dot-blot assay was developed for both PAb 1000/1201 and MAb 12.8. As shown in Fig. 22.1, EPS from *Mucor racemosus* gave the strongest immunoreaction with PAb 1000/1201 whereas *Rhizomucor pusillus* gave the highest density with MAb 12.8.

Table 22.1. Reactivity of MAb 12.8 and PAb 1000/1201 raised against the EPS of *Mucor racemosus* with EPS of different moulds (De Ruiter *et al.*, 1993b).

	Indirect ELISA titre*	
Mould strain	MAb 12.8	PAb 1000/1201
Mucor racemosus Fres., CBS 222.81	+++	+++
Mucor hiemalis Wehmer CBS 201.28	+++	++
Mucor circinelloides van Tieghem RIVM M 40	+	++
Rhizopus stolonifer (Ehrenb.) Lind CBS 609.82	++	++
Rhizopus oryzae Went and Prinsen Geerlings LU 581	++	++
Rhizomucor pusillus (Lindt) Schipper CBS 432.78	+++	++
Rhizomucor meihei (Cooney and Emerson) Schipper CBS 371.71	+++	++
Absidia corymbifera (Cohn) Sacc. and Trotter LU 017	+	++
Thamnidium elegans Link CBS 342.55	+	++
Syncephalastrum racemosum Cohn CBS 443.59	+	++
Mortierella reticulata van Tieghem CBS 452.74	−	−
Mortierella hyalina (Harz) W. Gams CBS 654.68	−	−
Mortierella polycephala Coemans CBS 327.72	−	−
Mortierella roseonana W. Gams CBS 473.74	++	++
Mortierella nana Linnem., CBS 730.70	++	++
Mortierella ramanniana (Möller) Linnem. var. *ramanniana* CBS 243.58	++	+
Mortierella ovata Yip CBS 499.82	+	+
Mortierella isabellina Oudem. CBS 560.63	+	+
Penicillium citrinum Thom CBS 117.64	−	−
Penicillium dierckxi Biourge RIVM M 90	−	−
Penicillium aurantiogriseum Dierckx CBS 342.51	−	−
Penicillium digitatum Sacc. RIVM M58	−	−
Aspergillus fumigatus Fres. RIVM M3	−	−
Aspergillus niger van Tieghem CBS 553.65	−	−
Fusarium solani (Mart.) Sacc. CBS 165.87	−	−
Fusarium poae (Peck) Wollenweber CBS 446.67	−	−
Alternaria alternata (Fr.) Keissler RIVM M 13	−	−
Botrytis cinerea Pers. RIVM M 17	−	−
Cladosporium cladosporioides (Fres.) de Vries CBS 143.65	−	−
Pichia membranaefaciens (Hansen) Hansen	−	++

Source: De Ruiter *et al.*, 1993b.
* Microtitre plates coated with 5 μg ml^{-1} of the respective EPS preparation. ELISA reactivity expressed as the titre, defined as the reciprocal dilution of a solution of 10 μg ml^{-1} antibody just giving a positive reaction. −, $\leq$ 10; +, 10 < titre $\leq$ 100; ++, 100 < titre $\leq$ 1000; +++, titre > 1000. Experiments were carried out in triplicate.

LIMITATIONS

MAb 12.8 raised against EPS isolated from *M. racemosus* was specific for all mould species belonging to the order of Mucorales except species of *Mortierella* subgenus *Mortierella*. No cross-reactions were observed with EPS from mould or yeast species tested other than Mucorales. A positive ELISA requires the presence of approximately 0.1 μg antigenic material coated to the wall of each well of the microtitre plate. An amount of 10 μg of antigenic material is needed for each spot in the dot-blot assay. This MAb was found to be

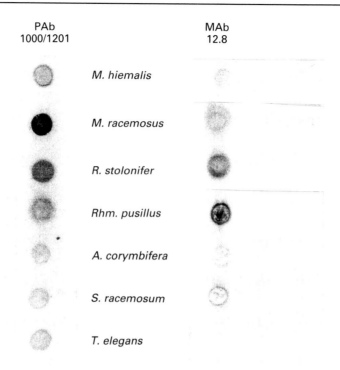

PAb   MAb
1000/1201   12.8

M. hiemalis

M. racemosus

R. stolonifer

Rhm. pusillus

A. corymbifera

S. racemosum

T. elegans

Fig 22.1.   Analysis with a dot-blot assay of various EPS isolated from species belonging to the order Mucorales with PAb 1000/1201 and MAb 12.8. After correction for the optical density of the nitrocellulose paper itself, the limits for black were set at OD 0.70 and 0.22 for PAb 1000/1201 and for MAb 12.8, respectively. (From De Ruiter *et al.*, 1993b, with permission from the Society for General Microbiology.)

slightly more specific for these moulds than the polyclonal antibodies raised against the same EPS (De Ruiter *et al.*, 1992b).

The immunological reaction of these fungal carbohydrate antigens with rabbit-raised polyclonal antibodies is based on recognition of 2-O-methyl-mannose containing epitopes (De Ruiter *et al.*, unpublished). The immunoreactivity of these EPS with MAb 12.8 is also based on carbohydrate epitopes. Preliminary partial acid hydrolysis experiments indicate that the fucose residues may play an important role in these epitopes (De Ruiter *et al.*, 1993b).

Apparently, different carbohydrate residues are immunoreactive with the rabbit-raised monospecific polyclonal and mouse-raised monoclonal antibodies. This can be attributed to either the different immunization procedures used or different host animal species used. Intrasplenic immunization was applied for raising MAb 12.8 in mice. It can be assumed that in this particular case the species difference might be the major factor for the difference in immune response.

Based on the results presented, it can be concluded that monoclonal antibodies are not necessarily more specific than polyclonal antibodies. It can be assumed that the specificity depends heavily on the presence of common and characteristic antigenic determinants in the polysaccharides.

With this MAb it will be feasible to test raw materials, food and feed on the presence of moulds belonging to the genera *Mucor, Rhizopus, Rhizomucor, Absidia, Thamnidium* and *Syncephalastrum* in industrial control laboratories and food inspection services (De Ruiter *et al.*, 1993a). Furthermore, in medicine these assays may be useful for diagnosis of mucormycosis in humans, by detection of circulating antigens (Kaufman *et al.*, 1989).

ACKNOWLEDGEMENTS

We gratefully acknowledge Dr W. Gams and Dr R. A. Samson (Centralbureau voor Schimmelcultures, Baarn, The Netherlands) for providing some of the strains of moulds used in this study and critical reading of the manuscript. The investigations were supported by The Netherlands' Foundation for Chemical Research (SON) with financial aid from The Netherlands' Technology Foundation (STW).

REFERENCES

Bos, E. S., Van der Doelen, A. A., Van Rooy, N. and Schuurs, A. H. W. M. (1981) 3,3',5,5'-tetramethylbenzidine as an Ames test negative chromogen for horseradish peroxidase in enzyme-immunoassay. *Journal of Immunoassay* 2, 187–203.

Bossi, R. and Dewey, F. M. (1992) Development of a monoclonal antibody-based immunodetection assay for *Botrytis cinerea*. *Plant Pathology* 41, 472–482.

De Ruiter, G. A., Van der Lugt, A. W., Voragen, A. G. J., Rombouts, F. M. and Notermans, S. H. W. (1991) High-performance size-exclusion chromatography and ELISA detection of extracellular polysaccharides from Mucorales. *Carbohydrate Research* 215, 47–57.

De Ruiter, G. A., Hoopman, T., Van der Lugt, A. W., Notermans, S. H. W. and Nout, M. J. R. (1992a) Immunochemical detection of Mucorales species in food. In: Samson, R. A., Hocking, A. D., Pitt, J. I. and King, A. D. (eds) *Modern Methods in Food Mycology.* Elsevier Publishers, Amsterdam, pp. 221–227.

De Ruiter, G. A., Van Bruggen-van der Lugt, A. W., Nout, M. J. R., Middelhoven, W. J., Soentoro, P. S. S., Notermans, S. H. W. and Rombouts, F. M. (1992b) Formation of antigenic extracellular polysaccharides by selected strains of *Mucor* spp., *Rhizopus* spp., *Rhizomucor* spp., *Absidia corymbifera* and *Syncephalastrum racemosum*. *Antonie van Leeuwenhoek* 62, 189–199.

De Ruiter, G. A., Notermans, S. H. W. and Rombouts, F. M. (1993a) Review: New methods in food mycology. *Trends in Food Science and Technology* 4, 91–97.

De Ruiter, G. A., Van Bruggen-van der Lugt, Bos, W., Notermans, S. H. W., Rombouts, F. M. and Hofstra, H. (1993b) The production and partial characterization of a monoclonal IgG antibody specific for moulds belonging to the order Mucorales. *Journal of General Microbiology* 139, 1557–1564.

De Ruiter, G. A., Van Bruggen-van der Lugt, A. W., Rombouts, F. M. and Gams, W. (1993c) Approaches to the classification of the *Mortierella isabellina* group (Mucorales): antigenic extracellular polysaccharides. *Mycological Research* 97, 690–696.

Dewey, F. M. (1992) Detection of plant-invading fungi by monoclonal antibodies. In: Duncan, J. M. and Torrance, L. (eds) *Techniques for the Rapid Detection of Plant Pathogens.* Blackwell Scientific, Oxford, pp. 47–62.

Dewey, F. M., MacDonald, M. M., Phillips, S. I. and Priestley, R. A. (1990) Development of monoclonal-antibody-ELISA and -DIP-stick immunoassays for *Penicillium islandicum* in rice grains. *Journal of General Microbiology* 136, 753–760.

Jarvis, B., Seiler, D. A. L., Ould, A. J. L. and Williams, A. P. (1983) Observations on the enumeration of moulds in food and feedingstuffs. *Journal of Applied Bacteriology* 55, 325–336.

Kaufman, L., Turner, L. F. and McLaughlin, D. W. (1989) Indirect enzyme-linked immunosorbent assay for zygomycosis. *Journal of Clinical Microbiology* 27, 1979–1982.

Notermans, S. and Heuvelman, C. J. (1985) Immunological detection of moulds in food by using the enzyme-linked immunosorbent assay (ELISA); preparation of antigens. *International Journal of Food Microbiology* 2, 247–258.

Pitt, J. I. and Hocking, A. D. (1985). *Fungi and Food Spoilage*. Academic Press, Sydney, 410 pp.

Samson, R. A. and Van Reenen-Hoekstra, E. S. (1988) *Introduction to Food-borne Fungi*. Centralbureau voor Schimmelcultures, Baarn, 299pp.

Stynen, D., Sarfati, J., Goris, A., Prévost, M. C., Lesourd, M., Kamphuis, H., Darras, V. and Latgé, J.-P. (1992) Rat monoclonal antibodies against *Aspergillus galactomannan*. *Infection and Immunity* 60, 2237–2245.

Sugar, A. M. (1992) Mucormycosis. *Clinical Infectious Disease* 14 (Suppl 1), S126–129.

23 Monoclonal Antibody-based ELISA for Detection of Mycelial Antigens of *Botrytis cinerea* in Fruits and Vegetables

R. Bossi,[1] L. Cole,[2] A. D. Spier[1] and F. M. Dewey[1]

[1] *Department of Plant Sciences, University of Oxford, South Parks Road, Oxford OX1 3RB, UK;* [2] *Oxford Brookes University, Gipsy Lane Campus, Headington, Oxford OX3 0BP, UK.*

BACKGROUND

Botrytis cinerea Pers.:Fr. [Teleomorph: *Botryotinia fuckeliana* (de Bary) Whetzel] is responsible for considerable losses of soft fruits, vegetables and cut flowers (Smith *et al.*, 1986; Salinas *et al.*, 1989). It is a common and serious disease of glasshouse crops that is known to cause 'brown rot' of stored vegetables such as cabbage, artichoke, onion, potato, beet and carrots (Jarvis, 1977; Coley-Smith *et al.*, 1980; Malathrakis, 1982; Locke and Fletcher, 1988).

Commonly infected fruits are strawberry, raspberry, tomato, sweet pepper, apples, pears and grapes. Infection frequently takes place at blossom time through the flowers but the fungus remains latent or quiescent until the fruit ripens (Dashwood and Fox 1988). Colonization of such fruits generally occurs postharvest when the fruit is *en route* to the market place (Sommer, 1982). Detection of the fungus at the latent stage, before packaging for transit, is difficult by classical methods and is time consuming. Relatively quick, user-friendly assays are needed to detect latent infections.

Grey mould of grapevine is probably the most significant disease caused by *B. cinerea*. Any herbaceous part of the grapevine is susceptible to attack but the flower heads and grape bunches are particularly prone. After flowering, the withered flowers tend to be caught in the developing bunches and serve as a saprophytic base for subsequent infection. Grey mould alters the quality of the wine giving taints and poor vinification properties, which makes the wine unstable and unsuitable for long-term maturation. The price paid for the grapes is based on the quality of the grape clusters and berries which is done currently by visual inspections. These inspections are generally inadequate and cause delays. A quantitative assay that could be performed on the grape juice after it has been expressed would be helpful (Ricker *et al.*, 1991).

Assays employing polyclonal antibodies raised in rabbits have recently been developed to detect *B. cinerea* in foods by Cousin *et al.* (1990) and in expressed grape juice by Ricker *et al.* (1991). However, the required specificity of the antiserum, which Ricker

165

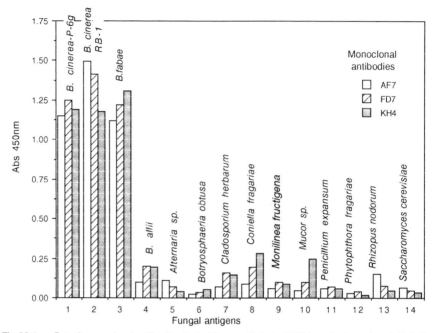

Fig 23.1. *Botrytis* monoclonal antibodies tested for specificity by ELISA against species of *Botrytis* and other fungi.

et al. (1991) achieved by using early bleeds, may not be easily reproducible. Furthermore, when they used antiserum from later bleeds they encountered considerable cross-reaction with *Aspergillus niger*. Du Pont have developed a polyclonal assay for the

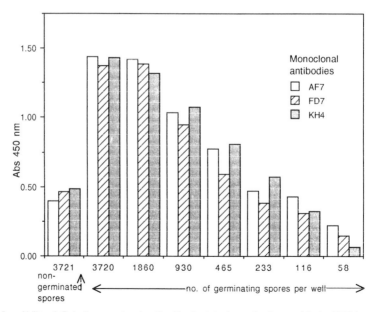

Fig 23.2. Ability of *Botrytis* monoclonal antibodies to detect germinating conidia by ELISA.

detection of *B. cinerea* in grape juice but only limited information is available on the specificity of their assay system (Cagnieul and Majarian, 1991).

We have developed a general-purpose assay for the detection of *B. cinerea* in expressed juice from infected strawberries and grapes.

Production of *B. cinerea*-specific monoclonal antibodies

Antigens for immunization were derived from surface washings of a single spore isolate of *B. cinerea* freshly isolated from grapes in Switzerland by Pezet (1990). The fungus was grown on potato dextrose agar in Petri dishes for 17 days at 21°C and washed with a total of 5 ml PBS. The combined cell-free washing fluid, which contained $<0.7\mu g\,ml^{-1}$ of protein, was passed through a centricon 30 kD filter (Amicon no. 4208) to remove the high molecular weight carbohydrates, proteins and glycoproteins. The low M_r fraction was retained, freeze dried and redissolved in 1 ml of distilled H_2O for use as the immunogen. Three 6-week-old female Balb/c mice were given four intraperitoneal injections of the immunogen at 2 week intervals. Each injection contained $300\mu l$ of the immunogen.

Hybridoma cells were produced by fusion of splenocytes, taken from a mouse 4 days after the first injection, with myeloma cells from an SP2/0-Ag 14 cell line by centrifugation in the presence of PEG M_r 1500 after the method of Kennet *et al.* (for details see Bossi and Dewey, 1992). Hybridoma supernatants were screened by ELISA, using antigen-coated wells, initially against PBS surface washings of the original isolate and later against surface washing and mycelial antigens of other fungi.

Three hybridoma cell lines were selected (BC-AF7, BC-FD7 and BC-KH4) that secreted antibodies that specifically recognize, by ELISA and immunofluorescence, *B. cinerea* and *B. fabae* but not apparently *B. allii* or other fungi normally involved in postharvest spoilage of fruits and vegetables (Fig. 23.1). Antibodies from one of these cell lines (BC-KH4) have been used to develop the antigen-based ELISA test, described later, for the detection of *B. cinerea* antigens in expressed juice from strawberries and grapes. Recognition of non-germinating spores by supernatants from all three cell lines is poor, but conidia that have been allowed to germinate overnight in microtitre wells or on glass slides can be detected by ELISA or immunofluorescence respectively (Fig. 23.2). The BC-KH4 antigen is known to be localized in the cell wall (Fig. 23.3(a) and (b)). Supernatants from BC-KH4 gave the lowest background values with healthy tissue. Indirect evidence from heat, protease and periodate treatment of the antigens indicates that the monoclonal antibody BC-KH4 recognizes carbohydrate epitopes on a glycoprotein. All three antibodies belong to the immunoglobulin subclass IgM. (For details see Bossi and Dewey, 1992; Dewey and Bossi, 1992).

PROTOCOL

ELISA – protocol

Materials

1. Multiwell polystyrene strips, and strip holder frames (Microstrip PS, ICN Biochemicals Ltd).

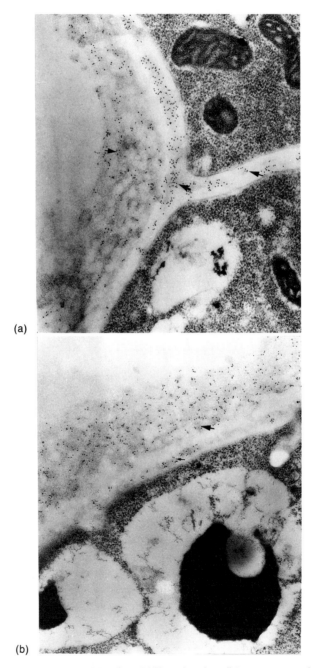

(a)

(b)

Fig 23.3.   Immunolocalization of the *Botrytis*-KH4 antigen in walls' cross-septa and extracellular matrix (arrows) of germlings of *B. cinerea*. Material was embedded in LR white and immunostained with 10 nm gold labelled secondary antibody.

2. Antigen extraction buffer – phosphate buffered saline (PBS), pH 7.2.
3. Washing buffer – PBS with 0.05% Tween 20.
4. Goat antimouse IgG + IgM horseradish peroxidase conjugate (Sigma 0412).
5. Substrate buffer: 5 ml 0.2 M sodium acetate, 195 μl 0.2 M citric acid, 5 μl 30% H_2O_2, 5 ml H_2O and 100 μl TMB stock substrate solution (10 mg ml^{-1} 3,3,5,5-tetramethyl-benzidine in dimethyl sulphoxide, stored at 4°C).
6. Test material.

Method

1. Squash the test plant material in a polythene bag with the extraction buffer (PBS) in a 1:5 ratio (w/v).
2. Centrifuge the extract at 2000 g for 5 min to remove plant debris and dilute with further PBS to give a final dilution of 1:20 (w/v).
3. Coat microtitre wells with the plant extract, 100 μl per well.
4. After incubating the plate overnight at 4°C invert and flick out residual antigen solution.
5. Wash four times with the washing buffer (PBST), once with PBS and briefly with water, dry in a laminar flow hood and store sealed at 4°C until ready for processing.
6. Block plate by incubating with PBS containing 0.3% casein for 30 min, 100 μl per well.
7. Wash plate four times with PBST.
8. Incubate with *Botrytis*-specific monoclonal antibody – BC-KH4 for 1 h, 100 μl per well.
9. Wash plate four times with PBST.
10. Incubate with secondary antibody-peroxidase (Sigma 0412 diluted 1 in 1000 in PBST), 100 μl per well.
11. Wash four times with PBST.
12. Add 100 μl of substrate solution to each well, incubate for 30 min.
13. Stop reaction by addition of 100 μl 3 M H_2SO_4 to each well.
14. Read absorbance at 450 nm on an ELISA-plate reader.

RESULTS

Relatively high absorbance values were obtained with extracts from strawberries that were naturally infected with *B. cinerea* but not with extracts from healthy strawberries or strawberries infected with *Mucor* sp. (Fig. 23.4). Using this assay system we can also detect and quantify *B. cinerea* antigens in the expressed juice from naturally and artificially infected grapes (Fig. 23.5). The sensitivity of the assay can be increased approximately 15-fold by using a biotinylated secondary antibody and a streptavidin conjugate (Fig. 23.5). Using this system we can detect the equivalent of 125 ng ml^{-1} of freeze dried *B. cinerea* mycelium.

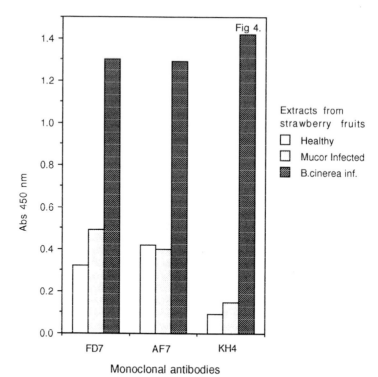

Fig 23.4. *Botrytis* monoclonal antibodies tested by ELISA against extracts from healthy strawberries and strawberries naturally infected, separately, with *B. cinerea* and *Mucor* sp.

LIMITATIONS

Non-specific binding of the commercial secondary antibody-peroxidase conjugate alone to wells coated with extracts from *Botrytis*-infected fruits but not extracts from healthy tissue has been noticed and is the subject of ongoing investigations. This non-specific binding has not been found in assays with extracts from tissues infected with other fungi; it can be reduced by including 0.3% casein in the conjugant diluent.

AVAILABILITY OF THE *BOTRYTIS*-SPECIFIC MONOCLONAL ANTIBODY BC-KH4

Contact F. M. Dewey, Department of Plant Sciences, University of Oxford, South Parks Road, Oxford OX1 3RB. All three cell lines are deposited with the European Collection of Animal Cell Cultures at the Public Health Laboratory, Porton Down, Salisbury, Wiltshire.

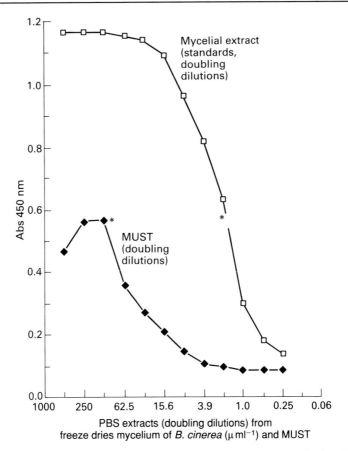

Fig 23.5. Concentration of *Botrytis* antigens in MUST, from an English winery, October 1992, determined by ELISA by using doubling dilutions in PBS together with extracts from freeze dried mycelium of *B. cinerea* as standards and the *Botrytis*-specific monoclonal antibody BC-KH4.

REFERENCES

Bossi, R. and Dewey, F. M. (1992) Development of a monoclonal antibody-based immunodetection assay for *Botrytis cinerea*. *Plant Pathology* 41, 472–482.

Cagnieul, P. and Majarian, W. (1991) Diagnostic kit for grey mould (*Botrytis cinerea*). In: *Proceedings of Third International Conference on Plant Protection, Bordeaux*, Dec. 1991, pp. 555–561.

Coley-Smith, J. R., Verhoeff, K. and Jarvis, W. R. (1980) *The Biology of* Botrytis. Academic Press, London.

Cousin, M. A., Dufrenne, J., Rombouts, F. M. and Notermans, S. (1990) Immunological detection of *Botrytis* and *Monoascus* species in food. *Food Microbiology* 7, 227–235.

Dashwood, E. P. and Fox, R. A. (1988) Infection of flowers and fruits of red raspberry by *Botrytis cinerea*. *Plant Pathology* 37, 423–430.

Dewey, F. M. and Bossi, R. (1992) Assay for *Botrytis cinerea*. UK Patent application filed by ISIS innovation Ltd. No. 9216305.4.

Elad, Y. and Volpin, H. (1991) Heat treatment for control of rose and carnation grey mould (*Botrytis cinerea*). *Plant Pathology* 40, 278–287.

Jarvis, W. R. (1977) Botryotinia *and* Botrytis *Species: Taxonomy, Physiology and Pathogenicity*. Monograph 15, Research Branch, Canada Department of Agriculture, Ottawa.

Locke, T. and Fletcher, J. T. (1988) Incidence of benomyl and iprodine resistance in isolates of *B. cinerea* in tomato crops in England and Wales in 1986. *Plant Pathology* 37, 381–384.

Malathrakis, N. E. (1982) Resistance of *Botrytis cinerea* to dichlofluanid in greenhouse vegetables. *Plant Diseases* 73, 138–141.

Pezet (1990) Federal Agricultural Research Station, Nyon, Switzerland.

Ricker, R. W., Marois, J. J., Dlott, J. W., Bostock, R. M. and Morrison, J. C. (1991) Immunodetection and quantification of *Botrytis cinerea* on harvested wine grapes. *Phytopathology* 81, 404–411.

Salinas, J., Glandorf, D. C. M., Picavet, F. D. and Verhoff, K. (1989) Effects of temperature, relative humidity and age of conidia on the incidence of spotting on gerbera flowers caused by *Botrytis cinerea*. *Netherlands Journal of Plant Pathology* 95, 51–64.

Sommer, N. F. (1982) Post harvest practices and post harvest disease of fruit. *Plant Disease* 66, 357–364.

Smith, I. M., Dunez, J., Phillips, D. H., Lelliott, R. A. and Archer, S. A. (eds) (1986) *European Handbook of Plant Disease*. Blackwell Scientific, Oxford, pp. 432–435.

24 Monoclonal Antibodies for Detection of Conidia of *Botrytis cinerea* on Cut Flowers Using an Immunofluorescence Assay

J. Salinas, G. Schober and A. Schots

Laboratory for Monoclonal Antibodies, PO Box 9060, 6700 GW Wagenigen, The Netherlands.

BACKGROUND

Botrytis cinerea Pers.:Fr. is an airborne plant-pathogenic fungus which causes considerable losses in many crops throughout the world. Jarvis (1980) reported more than 200 host plants for *B. cinerea*, most of them economically important, such as ornamental flowers and vegetables, grown both in fields and in glasshouses. Bulbs and fruits, including strawberries and grapes and even tree seedlings can be infected by the fungus. Host attack can occur before harvest or later during transport or storage. The fungus forms both conidia and microconidia as well as sclerotia. Infection by means of sclerotia is rarely observed. Mycelium from infected tissues and conidia produced on dead or dying plants are the most important sources of infection (Verhoeff, 1980).

Botrytis cinerea has become one of the most important threats to the production and export of cut flowers. Conidia produced in large amounts in greenhouses are easily spread through the air. After landing on the flowers the conidia remain dormant until water, for instance through condensation, is available for the spores to germinate and in this way flowers become infected within a few hours (Salinas *et al.*, 1989). Flowers auctioned in The Netherlands are sold in many countries. During transport to these countries they are exposed to varying temperatures which increases the chances of condensation. Therefore, it is important to ensure that flowers remain free of conidia.

The presently used tests for detection of *B. cinerea* are based on plating techniques, e.g. using selective media (Kerssies, 1990) or on immunoassays using polyclonal antibodies (Cousin *et al.*, 1990; Cagnieul and Majarian, 1991; Ricker *et al.*, 1991). The disadvantages of plating techniques are the time required and the necessity for some taxonomic skills. The major disadvantage of polyclonal antibodies is the lack of specificity. Monoclonal antibodies provide an alternative which enables the development of a sensitive, rapid and accurate detection method for *B. cinerea* (Bossi and Dewey, 1992; Salinas and Schots, 1992).

Monoclonal antibodies against conidia of B. *cinerea*

Three mouse monoclonal antibodies have been selected (Salinas and Schots, 1993). MAb 4H10 was raised against whole conidia, MAb 9E11 against extracellular proteins and MAb 14E5 against an esterase that is constitutively present (Salinas, 1992). 4H10 is an IgM, 9E11 and 14E5 are IgG_1 antibodies. The three MAbs bind to conidia of all of the 43 *B. cinerea* isolates that have been tested so far. They did not react with bacteria or spores from other fungi isolated from gerbera in greenhouses. Similarly no cross-reaction was observed with flower extracts. These MAbs also react with mycelium of *B. cinerea*.

The three MAbs are not species-specific because they also recognize conidia of *B. squamosa, B. elliptica, B. tullipae* and some *B. aclada* isolates. However, the possibility of false-positive reactions is very small because most of the *Botrytis* species are pathogenic to the flowers of interest (Salinas, unpublished results). All MAbs showed different fluorescence patterns with *B. cinerea* when compared to the other *Botrytis* species tested (Fig. 24.1).

Protocol

Materials

1. Liquid nitrogen, pestle and mortar.
2. Ultrasonic probe (UP-400, Sonicor Ltd).
3. Coating buffer: 5% horse serum in phosphate buffered saline with 0.05% Tween 20 (PBST).
4. Extraction buffer: 0.1% BSA in PBST (w/v).
5. Filter $45 \mu m$ pore size (Millipore Inc.).
6. Vacuum pump.
7. Horse serum (HyClone Laboratories Inc.).
8. Centrifuge.
9. Millipore MultiScreen-HV 96-well filtration plate, $0.45 \mu m$ pore size and MultiScreen Vacuum Manifold (Millipore Inc.).
10. 9E11, 14E5 and 4H10 MAbs–FITC (fluorescein isothiocyanate) conjugates in extraction buffer ($10 \mu g\, ml^{-1}$; w/v).
11. Antiquench stock solution: dissolve 100 mg *para*-phenylenediamine (Merck) in 10 ml PBS add 90 ml glycerol. Adjust pH to 8 with 0.5 M sodium carbonate buffer pH 8. Store at $-20°C$.

Method (see also Fig. 24.2)

1. Coat MultiScreen-HV filtration plate with $200 \mu l$ coating buffer per well at least 30 min at 37°C or overnight at 4°C.
2. Gerbera: sonicate four times (5 s each) 15 ray florets in 25 ml extraction buffer. Rose: crush rose petals in liquid nitrogen and add 25 ml extraction buffer.
3. Filtrate extracts (step 2) through a $45 \mu m$ pore size filter.
4. Centrifuge at 5000 g for 10 min to concentrate the conidia and resuspend the pellet in coating buffer.
5. Remove the contents of the MultiScreen plate wells with the MultiScreen Vacuum

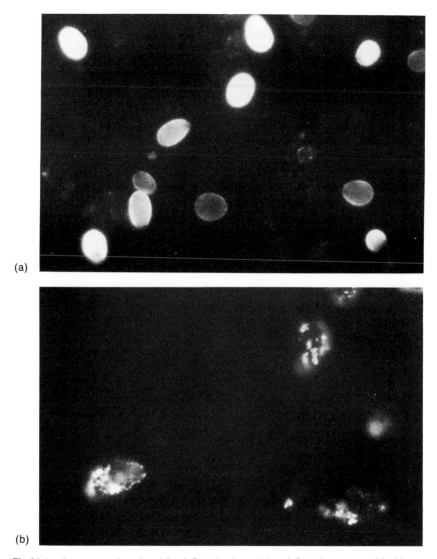

Fig 24.1.   Immunoreaction of conidia of *Botrytis cinerea* (a) and *Botrytis squamosa* (b) with antibodies 9E11 and 4H10 respectively. The conidia were visualized using a goat antimouse fluorescein isothiocyanate conjugate.

Manifold and incubate with conidia suspensions in the wells for 30 min at 20°C, 100 μl per well.

6. Rinse wells three times with PBST using the MultiScreen Vacuum Manifold.
7. Incubate the plates with 100 μl per well of MAb–FITC conjugate, for at least 30 min at 30°C in the dark.
8. Rinse conidia twice in PBST.
9. Incubate the conidia in ultra pure water for 5 min in the dark. Remove the water by vacuum.
10. Add 15 μl of antiquench per well.

COLLECTION OF CONIDIA

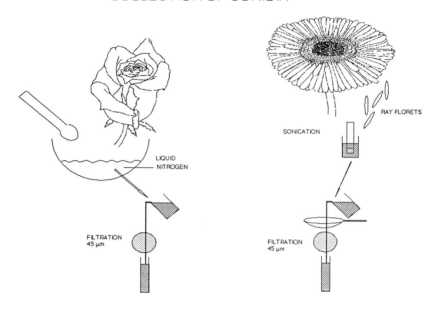

RAY FLORETS

SONICATION

LIQUID
NITROGEN

FILTRATION
45 μm

FILTRATION
45 μm

IMMUNOFLUORESCENCE TEST

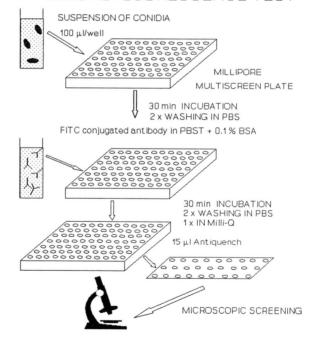

SUSPENSION OF CONIDIA

100 μl/well

MILLIPORE
MULTISCREEN PLATE

30 min INCUBATION
2 x WASHING IN PBS

FITC conjugated antibody in PBST + 0.1 % BSA

30 min INCUBATION
2 x WASHING IN PBS
1 x IN Milli-Q

15 μl Antiquench

MICROSCOPIC SCREENING

Fig 24.2. Detection of conidia of *Botrytis cinerea* present on gerbera and rose-infected flowers. Collection of conidia from the flowers by sonication of gerbera ray florets in PBS or crushing rose petals in liquid nitrogen (top). Visualization of conidia after immunoreaction with monoclonal antibodies 9E11 or 14E5 conjugated to fluorescein isothiocyanate (bottom).

11. Detach the conidia from the bottom filter of each well with a pipette.
12. Fluorescence microscopy examination of the conidia, on a 24 multiwell microscopy slide, placing 2 μl aliquots per well.

RESULTS AND LIMITATIONS

Botrytis cinerea conidia collected from field and laboratory-infected roses and gerberas reacted well in the test with the three selected MAbs (see Fig. 24.1(a) for the reaction with MAb 9E11). During the whole procedure some conidia will be lost or remain attached to the MultiScreen plate filters. This can be a limitation for testing rose flowers where inoculum concentrations of less than one conidia cm^{-2} have to be detected.

The assay is most sensitive if the reaction is visualized with a secondary antimouse–FITC labelled antibody. However, these secondary antibodies sometimes show 'non-specific' binding to other fungi. We have avoided this problem by using MAbs directly conjugated to FITC, although some sensitivity is lost. Fluorescence intensity can be improved without losing any specificity by extending the incubation time of the conidia with the MAb–FITC conjugates.

Availability of the *Botrytis*-specific monoclonal antibodies 9E11, 14E5 and 4H10

The three mentioned MAbs–FITC conjugates are available through the authors.

REFERENCES

Bossi, R. and Dewey, F. M. (1992) Development of a monoclonal antibody-based immunodetection assay for *Botrytis cinerea*. *Plant Pathology* 41, 472–482.

Cagnieul, P. and Majarian, W. (1991) Diagnostic kit for grey mold (*Botrytis cinerea*). In: *Proceedings of Third International Conference on Plant Protection*, Bordeaux, Dec. 1991, pp. 555–561.

Cousin, M. A., Dufrenne, J., Rombouts, F. M. and Notermans, S. (1990) Immunological detection of *Botrytis* and *Monascus* species in food. *Food Microbiology* 7, 227–235.

Jarvis, W. R. (1980) Taxonomy. In: Coley-Smith, J. R., Verhoeff, K. and Jarvis, W. R. (eds) *The Biology of* Botrytis. Academic Press, London, New York, Toronto, Sydney, San Francisco, pp. 1–18.

Kerssies, A. (1990) A selective medium for *Botrytis cinerea* to be used in a spore-trap. *Netherlands Journal of Plant Pathology* 96, 247–250.

Ricker, R. W., Marois, J. J., Dlott, J. W., Bostock, R. M. and Morrison, J. C. (1991). Immunodetection and quantification of *Botrytis cinerea* on harvested wine grapes. *Phytopathology* 81, 404–411.

Salinas, J. (1992) Function of Cutinolytic Enzymes in the Infection of Gerbera Flowers by *Botrytis cinerea*. Thesis, University of Utrecht, pp. 85–93.

Salinas, J. and Schots, A. (1992) Detection of *Botrytis cinerea* on gerbera flowers using monoclonal antibodies. In: Verhoeff, K., Malathrakis, N. E. and Williamson, B. (eds) *Recent Advances in* Botrytis *Research*, Pudoc Scientific Publishers, Wageningen, pp. 127–132.

Salinas, J. and Schots, A. (1993) Monoclonal antibodies for the detection of conidia of *Botrytis cinerea* on gerbera flowers, (manuscript submitted).

Salinas, J., Glandorf, D. C. M., Picavet, F. D. and Verhoeff, K. (1989) Effects of temperature, relative humidity and age of conidia on the incidence of spotting on gerbera flowers by *Botrytis cinerea. Netherlands Journal of Plant Pathology* 95, 51–64.

Verhoeff, K. (1980) The infection process and host–pathogen interactions, In: Coley-Smith, J. R., Verhoeff, K. and Jarvis, W. R. (eds) *The Biology of* Botrytis. Acadamic Press, London and New York, pp. 153–180.

25 Strawberry Blackspot Disease (*Colletotrichum acutatum*)

I. BARKER, G. BREWER, R. T. A. COOK, S. CROSSLEY AND S. FREEMAN

Central Science Laboratory, Ministry of Agriculture, Fisheries and Food, Hatching Green, Harpenden, Herts AL5 2BD, UK.

BACKGROUND

Strawberry blackspot is a destructive disease of strawberry (Anon., 1989), caused by *Colletotrichum acutatum*. The disease, a ripe fruit rot, occurs within the EEC (Directive 77/93/EEC Annexe II, A, II) but is not established within the UK, though it has been intercepted on imported planting material.

Currently, all consignments of strawberry planting material imported into the UK are sampled by MAFF and checked for the presence of *C. acutatum* by CSL Harpenden in order to protect the UK strawberry crop, valued at £72 million in 1990. The current diagnostic test involves treating strawberry petioles with paraquat which stimulates the fungus to develop and sporulate when it can be detected by observation under a low power microscope. This test requires a 7-day incubation period and assessment is labour intensive. A more rapid test for the detection and identification of *C. acutatum* on strawberry petioles would be more cost effective and reduce the time that planting material has to be held in storage prior to planting, while awaiting test results. Barker and Pitt (1988) demonstrated that *C. acutatum* could be detected in anemone corms by ELISA using a polyclonal antibody raised to the fungus. It was decided to extend this work to try and develop a rapid serological assay for the detection of *C. acutatum* on imported strawberry petioles. However, in the case of strawberry blackspot disease, it is also necessary to distinguish *C. acutatum* from the other closely related species of *Colletotrichum*, *C. gloeosporioides* (*Glomerella cingulata*) and *C. fragariae*, known to cause other diseases on strawberry (Howard *et al.*, 1992). It appeared likely that the production of monoclonal antibodies to *C. acutatum* would be necessary to achieve this level of specificity.

IMMUNOGEN PREPARATION AND PRODUCTION OF MONOCLONAL ANTIBODIES

A spore suspension from a culture of *C. acutatum*, isolated from strawberry and identified by IMI, was used to inoculate shake flasks containing 50 ml of Czapek Dox broth. Five-day-old cultures were harvested and centrifuged (3000 rpm) for 5 min and the supernatant discarded. The pellets were resuspended in sterile PBS and recentrifuged. After one further cycle of centrifugation, pellets were stored at −70°C. Pellets were thawed and 20 g of material were homogenized with 20 ml of cold sterile PBS using a mortar and pestle. Homogenates were diluted with PBS to give a final protein concentration of 3 mg ml^{-1} and re-frozen at −70°C in 1 ml aliquots.

Female LOU/OLA rats (Harlan-Olac) were immunized intramuscularly with 1.5 mg total protein, emulsified with Freund's complete adjuvant, followed 3 weeks later by an immunization with 1.5 mg total protein, intramuscularly, in Freund's incomplete adjuvant. Seven weeks after the first immunization the final boost of 1.5 mg of protein in sterile PBS was given intravenously. Splenocytes were then prepared from spleens taken from rats, 3 days after the final intravenous injection, and either fused immediately or frozen in liquid nitrogen for future use. Splenocytes were fused with Y3 Ag 1.2.3. myeloma cells (Galfré *et al.*, 1979). Hybridoma selection was carried out using azaserine–hypoxanthine medium.

PROTOCOL

Infection of strawberry material

1. *Infection in vitro.* Detached petioles of strawberry (cv. Elsanta) were lightly wounded, inoculated with agar plugs containing mycelium of *C. acutatum* or *C. gloeosporioides* and incubated for 7 days. Their sap was squeezed out with a roller and used in the ELISA test. Healthy petioles were used as controls.
2. *Infection in vivo.* Strawberry plants (cv. Elsanta) were sprayed with a conidial suspension of the standard *C. acutatum* isolate and incubated for at least 1 week in a glasshouse (minimum temperature 20°C).

ELISA screening assay

A modified indirect ELISA coating antigen on to the plate was used to screen hybridomas and to evaluate sensitivity and specificity of detection. Fungal test antigens consisted of freeze-dried conidia and mycelium scraped from agar Petri plate cultures and adjusted to 10 mg ml^{-1} diluted in 0.05 M carbonate buffer (pH 9.6). Strawberry plant material was tested at a final dilution of 1/50 (w/v), homogenized and diluted in carbonate buffer. A modification of the ELISA test (dot blot) involved the replacement of the ELISA plate with a nylon membrane (PALL-Biodyne) and the use of insoluble substrate products (BCIP and NBT). The protocol for the indirect ELISA screening method was as follows:

1. Coat 96 well (NUNC) plates with 100 μl antigen or plant sap extract.
2. Incubate overnight at 4°C.

3. Wash three times (PBST).
4. Add $100 \mu l$ hybridoma tissue culture supernatant (1/10) or part-purified ascites $(1/1000$ from a $1 \, mg \, ml^{-1}$ IgG stock) diluted in PBST + 0.2% BSA.
5. Incubate for 1–2 h at 30°C.
6. Wash three times (PBST)
7. Add $100 \, \mu l$ rabbit antirat IgG alkaline phosphatase conjugate diluted in PBST + 0.2% BSA according to manufacturers' instructions.
8. Incubate for 1–2 hours (30°C).
9. Wash three times (PBST).
10. Add $100 \, \mu l$ substrate solution (*p*-nitrophenol, $0.6 \, mg \, ml^{-1}$ in diethanolamine buffer, pH 9.8).
11. Incubate for 30 min at room temperature and read at 405 nm, blanking on substrate solution.

RESULTS AND INTERPRETATION

At least 15 out of 23 MAbs tested exhibited a high degree of specificity towards *C. acutatum* in pure culture when challenged with a range of fungal species. They consistently reacted with 28 out of 33 isolates known to belong to *C. acutatum* and with only 1 out of 19 isolates from typical isolates of the closely related *C. gloeosporoides*. They have given inconsistent results with a small group of isolates from ceanothus and lupin, which previous workers have referred to as *C. gloeosporioides*, but show some characteristics of *C. acutatum*. These antibodies did not react to isolates of *C. fragariae* (eight isolates), *C. malvarum*, *C. dematium*, *C. circinans* and *C. coccodes*. They also did not react with 12 isolates of other fungal genera including a range of strawberry pathogens and saprophytes.

One, less specific, MAb reacted with 66 out of 67 isolates belonging to the genus *Colletotrichum* indicating a level of specificity towards the whole genus. Weak cross-reaction with one isolate of *Gliocladium roseum* was noted at high antigen concentration.

Using strawberry material infected *in vitro*, the specific MAbs have consistently distinguished between petioles infected with *C. acutatum* and those infected with *C. gloeosporioides* (Table 25.1). With material infected *in vivo*, however, they detected only 8 positives out of 50 petioles infected with *C. acutatum* compared with over 30 positives

Table 25.1. ELISA testing of infected strawberry petioles.

	ELISA value (A 405 nm) MAb	
Petiole sap	25/164.32	25/214.31
C. acutatum	1.292	0.619
C. gloeosporioides (G. cingulata)	0.042	0.598
Healthy	0.021	0.062
Buffer	0.024	0.025

detected by plating out the sap on to a selective medium (Barker and Pitt, 1987) or treatment with paraquat. The MAbs failed to detect *C. acutatum* when fewer than eight conidia (viable propagules) were present in the extract.

By centrifuging fungal preparations it was shown that the antigen was highly soluble in water and its activity was directly proportional to the number of conidia present. Immunofluorescent microscopy also indicated that the *C. acutatum*-specific MAbs reacted to conidial antigens but not to hyphae or resting structures. The genus-specific MAbs appeared to detect both conidia and hyphae.

Specific monoclonal antibodies have been developed that can identify most pure cultures of *C. acutatum* within 24 h using an ELISA test and within 3 h using a dot blot test. The taxonomy of some isolates giving inconsistent results needs to be clarified using molecular methods such as PCR.

The current assay is not sensitive enough to detect latent infection in strawberry petioles, possibly because the fungus exists mainly as resting mycelium rather than conidia. A combination of the paraquat test to stimulate sporulation and an ELISA test to identify the resultant conidia could considerably reduce the test time and be less labour intensive. Such a test is currently being evaluated along with further MAb development work, aimed at identifying a *C. acutatum*-specific MAb capable of detecting mycelial structures. The 'genus-specific' MAb is also being evaluated for the detection of tropical fruit diseases caused by *C. gloeosporioides*.

ACKNOWLEDGEMENTS

The financial support of MAFF (Plant Health Division/CSG) is acknowledged.

REFERENCES

Anon. (1989) Strawberry blackspot. Campaign Leaflet No. 73, MAFF.

Barker, I. and Pitt, D. (1987) Selective medium for the isolation from soil of the leaf curl pathogen of anemones. *Transactions of the British Mycological Society* 88(4), 553–555.

Barker, I. and Pitt, D. (1988) Detection of the leaf curl pathogen of anemones in corms by enzyme-linked immunosorbent assay (ELISA). *Plant Pathology* 37, 417–422.

Galfré, G., Milstein, C. and Wright, B. (1979) Rat × rat hybrid myelomas and monoclonal anti-Fd portion of mouse IgG. *Nature, (London)* 277, 131–133.

Howard, C. M., Maas, J. L., Chandler, C. K. and Albregts, E. E. (1992) Anthracnose of strawberry caused by the *Colletotrichum* complex in Florida. *Plant Disease* 76(10), 976–981.

26 Detection of the Anthracnose Pathogen *Colletotrichum*

P. R. MILLS, S. SREENIVASAPRASAD AND A. E. BROWN

Biotechnology Centre for Animal and Plant Health, The Queen's University of Belfast, Newforge Lane, Belfast BT9 5PX, UK.

BACKGROUND

The genus *Colletotrichum* Corda contains an extremely diverse collection of fungi including both saprophytes and phytopathogens. To date there have been 900 'species' assigned to *Colletotrichum* which have been reduced recently to 39 accepted taxa (Sutton, 1992).

Colletotrichum contains a number of pathogens which cause serious losses in cereals, grasses, fruits, legumes, vegetables and perennial crops particularly in subtropical and tropical regions. Disease symptoms are frequently depressed black lesions termed anthracnose but leaf-blights are also common.

C. gloeosporioides is the causal agent of anthracnose on an extremely wide range of plants including legumes (Lenne, 1992) and temperate and tropical fruit crops (Mordue, 1967; Jeffries *et al.*, 1990). The species is extremely variable in its morphology (Simmonds, 1965; Sutton, 1992).

C. kahawe (= '*C. coffeanum*') (Waller *et al.*, 1993) causes coffee berry disease (CBD) in Africa (Masaba and Waller, 1992). *C. kahawe* is morphologically similar to *C. gloeosporioides* but has been distinguished by its ability to infect green berries (Waller, 1982) and by biochemical characteristics (Waller *et al.*, 1993).

C. acutatum infects a wide range of hosts including strawberry. Anthracnose (black-spot) of strawberry has been reported from most strawberry growing regions of the world and is common in Europe (Howard and Albregts, 1983; Maas and Howard, 1985; Cook and Popple, 1984; Harris, 1992). Severe infection of strawberry plants has been recorded in the USA (Howard and Albregts, 1983; Wilson *et al.*, 1990). In the UK, *C. acutatum* is regarded as a potential threat to the strawberry industry and detection of the disease results in the compulsory destruction of the crop (Anon., 1987). *C. acutatum* can occur on strawberry plants as cryptic infections, and sometimes in lesions in complexes with *C. gloeosporioides* and *C. fragariae*. The overlapping symptoms and difficulty in recognizing characteristics distinguishing these organisms has led to considerable uncertainty regarding the causal agent of various forms of strawberry anthracnose.

Correct identification of *C. gloeosporioides* and *C. acutatum* should be possible using spore characteristics. However, recently it has become apparent that some isolates of *C. acutatum* appear to be morphologically indistinguishable from *C. gloeosporioides*. As *C. acutatum* is a quarantine organism in the UK it is vital that this species can be distinguished from other endemic species.

   C. coccodes causes anthracnose of vegetables in the Solanaceae and Cucurbitaceae (Mordue, 1967). *C. coccodes* is especially destructive on tomatoes grown for processing.

PROTOCOL

Materials

   1. Perkin Elmer Thermal Cycler; Hybaid Thermal Cycler.
   2. Species-specific primer sequences were derived from the internally transcribed spacer (ITS) regions 1 and 2 of the ribosomal DNA (rDNA) repeat unit.
   3. ITS 4 primer sequence was as described by White *et al.* (1990). Oligonucleotides were synthesized by Operon Technologies, Inc., California.
   4. PCR buffer: 200 mM Tris-HCl, pH 7.5; 250 mM NaCl, 25 mM EDTA, 0.5% SDS.
   5. PCR buffer was used as extraction buffer for infected tomato fruit.
   6. Extraction buffer for infected strawberry leaves/stolons: 200 mM Tris-HCl, 250 mM NaCl, 25 mM EDTA, 0.5% SDS, pH 8.5.
   7. QBT: 750 mM NaCl, 50 mM 3-(N-morpholino) propanesulphonic acid (MOPS), 15% ethanol, pH 7.0, 0.15% Triton X-100.
   8. QC: 1.0 M NaCl, 50 mM MOPS, 15% ethanol, pH 7.0.
   9. QF: 1.25 M NaCl, 50 mM MOPS, 15% ethanol, pH 8.2.
   10. Qiagen-tips, Hybaid Ltd.

Method

   1. *Fungal culture.* Fungal mycelium was produced in liquid shake culture at 25°C in glucose yeast extract medium (glucose, 10 g; $NH_4H_2PO_4$, 1 g; KC1, 0.2 g; $MgSO_4.7H_2O$, 0.2 g; yeast extract (Difco), 5 g; 0.5% w/v $CuSO_4.5H_2O$, 1 ml; 1% w/v $ZnSO_4.7H_2O$, 1 ml in 1 l distilled water) on an orbital shaker (120 rpm).
   2. *DNA extraction from fungal mycelium.* Total DNA was extracted from freeze-dried mycelial powder (300 mg) using a modification of the method of Raeder and Broda (1985).
      (a) Grow fungal isolates in suitable liquid medium (avoid excess agar medium with the inoculum).
      (b) Harvest mycelium by filtration after 3–5 days' growth.
      (c) Freeze mycelium in liquid nitrogen (10 min), pulverize and freeze dry.
      (d) Grind the mycelium to a fine powder and store at −70°C until use.
      (e) For extracting the DNA, mix 300 mg of mycelial powder with 3 ml of extraction buffer (Tris-HCl 200 mM, NaCl 250 mM, EDTA 25 mM and SDS 0.5%, pH 8.5) and incubate at room temperature for 15 min.
      (f) Add 2.1 ml phenol (which has been equilibrated with 1 M Tris-HCl pH 8.5 (twice) and with the extraction buffer and stored at 4°C under a layer of extraction buffer) and 0.9 ml chloroform and mix well.

(g) Centrifuge at 13,000 rpm for 1 h.

(h) Remove the aqueous phase, add 100 μl of RNAase solution (20 mg ml⁻¹ stock) and incubate for 15 min at 37°C (RNAase buffer: Tris HCl 10 mM and NaCl 15 mM, pH 7.5).

(i) Repeat twice phenol: chloroform extraction (13,000 rpm, 30 min).

(j) Remove aqueous phase and extract with 3 ml of chloroform (13,000 rpm, 15 min).

(k) Remove aqueous phase, add 0.54 volume of isopropanol and mix well to precipitate DNA at room temperature.

(l) Centrifuge briefly to pellet DNA.

(m) Discard the solution, wash the pellet with 70% alcohol.

(n) Vacuum dry the pellet to remove traces of alcohol.

(o) Add appropriate volume (500–700 μl) of TE8 (Tris HCl 10 mM and EDTA 1 mM, pH 8) and dissolve DNA by gentle agitation (e.g. on a rotawheel).

3. *DNA extraction from infected tomato fruit.*

(a) Remove approximately 500 mg of tissue from the periphery of a lesion to a microcentrifuge tube.

(b) Add an equal volume of PCR buffer and grind with a tapered glass rod.

(c) Spin homogenate at 14,000 g for 1 min in a microcentrifuge.

(d) Remove supernatant to a clean microcentrifuge tube and to it add an equal volume of isopropanol.

(e) Incubate for 5 min at room temperature.

(f) Collect precipitated DNA by centrifugation at 14,000 g for 10 min.

(g) Discard supernatant and resuspend pellet in 100 μl of TE8.

4. *DNA extraction from infected strawberry tissue.*

(a) Remove approximately 500 mg of tissue and grind in liquid nitrogen in a pestle and mortar.

(b) Suspend the homogenate in 400 μl of extraction buffer (see Materials), vortex and leave at room temperature for 10 min.

(c) Spin homogenate at 14,000 g for 2 min in a microcentrifuge.

(d) Remove 300 μl of supernatant to a clean tube and add 300 μl of isopropanol.

(e) Incubate at room temperature for 5 min.

(f) Collect precipitated DNA by centrifugation at 14,000 g for 10 min.

(g) Discard supernatant and resuspend pellet in 100 μl of TE8.

(h) Precipitate 50 μl of extract with 2.5 volumes ethanol, 0.1 volume sodium acetate for 1 h (or overnight if appropriate).

(i) Collect precipitated DNA by centrifugation at 14,000 g for 10 min.

(j) Discard supernatant and redissolve pellet in 200 μl QBT buffer.

(k) Apply sample to QBT-equilibrated Qiagen-tip and allow to flow through resin under gravity.

(l) Wash the Qiagen-tip with 2 × 500 μl QC buffer.

(m) Elute DNA with 500 μl of QF buffer.

(n) Precipitate DNA with 0.5 volume of isopropanol and collect pellet as above.

(o) Wash pellet with 70% ethanol, vacuum dry and redissolve in 100 μl of TE8.

5. *PCR amplification using target primers and product analysis.* Species-specific target primers were synthesized for *C. gloeosporioides* (CgInt), *C. acutatum* (CaInt) and *C. coccodes* (CcInt1) from the internally transcribed spacer (ITS) region 1 of the rDNA gene

block. A second primer was synthesized for *C. coccodes* (CcInt2) from the ITS 2 region. Target primers CgInt and CaInt were used with the conserved primer ITS 4 which has a sequence common to most (if not all) fungi. CcInt1 was used with CcInt2.

(a)   Set up PCR reaction mixtures (100 μl) to contain: 10 × *Taq* buffer, 10 μl; dATP, dCTP, dGTP, dTTP, 200 μM, 16 μl; 1 μM target primer, 5 μl; 1 μM ITS4 (or target primer 2), 5 μl; sterile distilled water, 53.5 μl; DNA (approx 50 ng), 10 μl; *Taq* DNA polymerase (2.5 U), 0.5 μl.

(b)   Programme thermal cycler for 30 cycles of 1.5 min at 94°C, 2 min at appropriate annealing temperature (determined for each target primer according to sequence, i.e. 2°C for each A and T and 4°C for each G and C) and 3 min at 72°C.

(c)   Following amplification, separate PCR products (10–15 μl) by electrophoresis in 1.4% (w/v) agarose gels containing ethidium bromide (0.4 μg ml^{-1}) and view on a UV transilluminator.

RESULTS AND LIMITATIONS

C. gloeosporioides

Specificity of C. gloeosporioides *target primer CgInt*

Nucleotide sequences of the ITS 1 region of several *C. gloeosporioides* isolates were aligned with the same region of other *Colletotrichum* spp. A 20 base region specific to *C. gloeosporioides* with little or no homology was identified, and used in the design of the internal primer CgInt (Fig. 26.1).

Primer CgInt and primer ITS 4 were used in PCR with genomic DNA from a range of isolates of *C. gloeosporioides*, *C. fragariae*, *C. kahawe*, *C. acutatum*, *C. lindemuthianum*, *C. orbiculare*, *C. linicola*, *C. capsici*, *Trichoderma* spp. and *Nectria galligena*. A fragment of approximately 450 bp was amplified with all *C. gloeosporioides* isolates (25), *C. fragariae* isolates (9) and *C. kahawe* isolates (10) tested. No amplification of the product was observed with DNA from any other fungi (Fig. 26.2).

```
C. gloeosporioides    GGCCTC**CCGCCTCCG*GGCGG
      C. kahawae      ------**-----C---*-----
     C. fragariae     ------**-----C--CC-----
     C. fragariae     ------**-----C---*-----
        C. musae      ----C-**-----C--AG----*
      C. acutatum     ---GC-GG--C--A--AC--G-A
      C. acutatum     ---GC-GG--C-*GT-AC--G--
       C. capsici     -T-TC-**-G---CT-TC**-C-
 C. lindemuthianum    **--C-**G----CG-TC-----
         C. lini      ---TC-GGG--T-C****-----
      C. orbiculare   -CT-C-GG*******TAAAAG--
```

Fig 26.1.   Variable region in the ITS 1 sequence of *Colletotrichum gloeosporioides* aligned with other *Colletotrichum* species.

Amplification of the C. gloeosporioides-*specific fragment from DNA from infected tissue*

A product of identical size (450 bp) to that amplified from fungal DNA was produced when primers CgInt and ITS 4 were used in PCR with total nucleic acid extracted from infected tomato tissue. No amplification product was observed with nucleic acid extracted from healthy tissue. Hybridization analysis of the PCR product from infected tissue using the PCR product from fungal DNA as a probe yielded a strong signal (Mills *et al.*, 1992). No signal was obtained with DNA from healthy tomato tissue. This confirmed that the product amplified contained the target sequence of the fungal DNA from infected tissue.

Sensitivity of detection

The sensitivity of PCR detection was assessed using serial dilutions of a known quantity of *C. gloeosporioides* genomic DNA as template. Successful amplification was achieved with 10 fg of fungal DNA (Mills *et al.*, 1992).

C. acutatum

Specificity of C. acutatum *target primer CaInt*

Nucleotide sequences of the ITS 1 region of 12 *C. acutatum* isolates were aligned with the same region from other *Colletotrichum* species. A 17-base region specific to *C. acutatum*, with little or no homology to other *Colletotrichum* species was identified and used in the design of the internal primer CaInt.

Primers CaInt and ITS 4 were used in PCR with genomic DNA from a range of *C. acutatum* isolates, three isolates each of *C. gloeosporioides* and *C. fragariae* as well as an isolate each of a number of other *Colletotrichum* species. A fragment of approximately 450 bp was amplified with the 10 *C. acutatum* isolates tested. One isolate each of *C. gloeosporioides* and

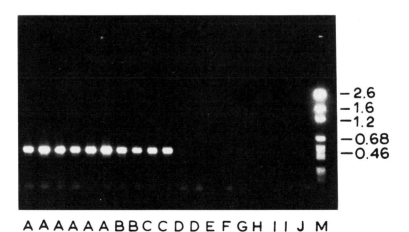

Fig 26.2. Amplification of a specific fragment from fungal DNA using target primers CgInt and ITS4 (A, *Colletotrichum gloeosporioides*; B, *C. kahawe*; C, *C. fragariae*; D, *C. acutatum*; E, *C. lindemuthianum*; F, *C. orbiculare*; G, *C. linicola*; H, *C. capsici*; I, *Trichoderma* spp.; and J, *Nectria galligena*). M, digested pGEM used as molecular size marker.

C. fragariae showed cross-reaction. The PCR conditions, including the possible use of a second specific primer, are being optimized to achieve greater specificity.

C. coccodes

Primer CcInt1 was used with CcInt2 to produce a product of approximately 350 bp. No product was amplified from any other *Colletotrichum* spp. or members of other genera tested.

A fragment of identical size (350 bp) to that amplified from fungal DNA was produced when primers CcInt1 and CcInt2 were used in PCR with total nucleic acid extracted from infected tomato root and/or fruit tissue.

Development of sensitive and preferably rapid diagnostic tests would assist in the identification of certain *Colletotrichum* spp., particularly those occurring in complexes with related species.

This study has shown that species-specific PCR primers can be derived from the internally transcribed spacer regions of the rDNA repeat unit. These regions have advantages over other parts of the genome as they occur in high copy number (at least 100 copies per genome) and are always present. These features ensure reliability and sensitivity. Within the genus *Colletotrichum* it would appear that species-specific primers could be developed for most, if not all, species. Primer sequences of very closely related taxa may differ by only two or three bases but PCR conditions could be optimized to permit differentiation. This is exemplified by our ability to separate American and European isolates of *C. acutatum* using a primer which differs by 3–4 bases between the two groups (data not presented).

Primers designed for diagnosis of *C. coccodes* did not react with any other *Colletotrichum* spp. The primer CgInt reacted with *C. gloeosporioides*, *C. fragariae* and *C. kahawe* suggesting a degree of cross-reactivity. However, other molecular data involving RFLP analyses of ribosomal and mitochondrial DNA and sequence analysis suggest that both *C. fragariae* and *C. kahawe* fall within the acceptable range of variation described for *C. gloeosporioides* and should not be considered as separate species (Sreenivasaprasad *et al.*, 1993a,b).

Primer CaInt cross-reacted with both an isolate of *C. gloeosporioides* and *C. fragariae*. A second species-specific primer from the ITS 2 region will probably eliminate this cross-reactivity.

In this study *C. gloeosporioides* and *C. coccodes* were successfully detected in infected tomato tissue. Amplification of the *C. gloeosporioides*-specific product from femtogram quantities of fungal DNA indicates the potential sensitivity of such a test. Crude estimates of mycelial quantities based on DNA yields suggest that 10 fg of DNA could represent as little as 100 pg of mycelium. This has obvious implications for detection of low level or quiescent infections.

However, detection of *C. acutatum* in DNA extracted from infected strawberry plants required additional purification steps involving passage through ion exchange chromatography tips. This step was thought to be necessary to remove unidentified inhibitors of the PCR reaction. The presence of inhibitors in plant tissues may also reduce the sensitivity of a PCR test.

At present, diagnosis of fungal pathogens using PCR would be unsuited to handling large sample numbers. A trained operator may be limited to 50–60 samples per day.

However, the sensitivity of the test allows sample bulking. Furthermore, a low level of expertise is required by the sample manipulations and the interpretation of results.

REFERENCES

Anon. (1987) *Plant Health Order* (Gt Britain).

Cook, R. T. A. and Popple, S. C. (1984) Strawberry black spot caused by *Colletotrichum* sp. (Abstract). Agricultural Division Advisory Service Plant Pathology Technical Conference, Harrogate, England.

Harris, D. (1992) A matter of good breeding. *The Grower* 5 March, 27–33.

Howard, C. M. and Albregts, E. E. (1983) Black leaf spot phase of strawberry anthracnose caused by *Colletotrichum gloeosporioides*. *Plant Disease* 67, 1144–1146.

Jeffries, P., Dodd, J. C., Jeger, M. J. and Plumbley, R. A. (1990) The biology and control of *Colletotrichum* species on tropical fruit crops. *Plant Pathology* 39, 343–366.

Lenne, J. M. (1992) *Colletotrichum* diseases of legumes. In: Bailey, J. A. and Jeger M. J. (eds) Colletotrichum – *Biology, Pathology and Control*. CAB International, Wallingford, pp. 237–249.

Maas, J. L. and Howard, C. M. (1985) Variation of several anthracnose fungi in virulence to strawberry and apple. *Plant Disease* 69, 164–166.

Masaba, D. and Waller, J. M. (1992) Coffee berry disease: The current status. In: Bailey, J. A. and Jeger, M. J. (eds) Colletotrichum – *Biology, Pathology and Control*. CAB International, Wallingford, pp. 237–249.

Mills, P. R., Sreenivasaprasad, S. and Brown, A. E. (1992) Detection and differentiation of *Colletotrichum gloeosporioides* isolates using PCR. *FEMS Microbiology Letters* 98, 137–144.

Mordue, J. E. M. (1967) Colletotrichum coccodes. *Descriptions of Pathogenic Fungi and Bacteria*, No. 131. Commonwealth Mycological Institute, Kew, Surrey, England.

Raeder, U. and Broda, P. (1985) Rapid preparation of DNA from filamentous fungi. *Letters in Applied Microbiology* 1, 17–20.

Simmonds, J. H. (1965) A study of the species of *Colletotrichum* causing ripe fruit rots in Queensland. *Journal of Agriculture and Animal Science* 22, 437–459.

Sreenivasaprasad, S., Brown, A. E. and Mills, P. R. (1993a) DNA sequence variation and interrelationships among *Colletotrichum* species causing strawberry anthracnose. *Physiological and Molecular Plant Pathology* 41, 265–281.

Sreenivasaprasad, S., Brown, A. E. and Mills, P. R. (1993b) Coffee berry disease pathogen in Africa: genetic structure and relationship to the group species *Colletotrichum gloeosporioides*. *Mycological Research* 97, 995–1000.

Sutton, B. C. (1992) The genus *Glomerella* and its anamorph *Colletotrichum*. In: Bailey, J. A. and Jeger, M. J. (eds) Colletotrichum – *Biology, Pathology and Control*. CAB International, Wallingford, pp. 1–26.

Waller, J. M. (1982) Some mycological aspects of the coffee berry disease pathogen. *Proceedings of the 1st Regional Workshop "Coffee Berry Disease"*, Addis Ababa, pp. 125–130.

Waller, J. M., Bridge, P. D., Black, R. L. and Hakiza, G. (1993) Differentiation of the coffee berry disease pathogen. *Mycological Research* 97, 989–994.

White, T. J., Bruns, T., Lee, S. and Taylor, J. (1990) Amplification and direct sequencing of fungal ribosomal RNA genes for phytogenetics. In: Innis, M. A., Gelfand, D. H., Sninsky, J. J. and White, T. J. (eds) *A Guide to Methods and Applications*. Academic Press, San Diego, pp. 315–322.

Wilson, L. L., Madden, L. J. and Ellis, M. A. (1990) Influence of temperature and wetness duration on infection of immature and mature strawberry fruit by *Colletotrichum acutatum*. *Phytopathology* 80, 111–116.

27 Development of Species-specific Probes for Identification of *Pyrenophora graminea* and *P. teres* by Dot-blot or RFLP

K. HUSTED

Research Centre for Plant Protection, Danish Institute of Plant and Soil Science, Lottenbergvej 2, DK-2800 Lyngby, Denmark.

BACKGROUND

The fungi *Pyrenophora graminea* Ito and Kurib. (anamorph *Drechslera graminea* (Rabenh. ex Schlecht.) Shoem.,) and *Pyrenophora teres* Drechs. (anamorph *Drechslera teres* (Sacc.) Shoem.), are serious seed-borne pathogens of barley. The two fungi are closely related (Smedegård-Petersen, 1976), but cause different diseases. *P. teres* is the causal agent of net blotch and leaf spot disease. *P. teres* Drechs. f. *teres* induces net-like necrotic lesions on the leaves while *P. teres* Drechs. f. *maculata* Smedeg. induces brown elliptical leaf lesions (Smedegård-Petersen, 1971). *P. graminea* is a systemic fungus which causes leaf stripe disease. In contrast to *P. teres*, *P. graminea* is not capable of infecting barley leaves directly and thus cannot cause epidemic infections (Smedegård-Petersen, 1971, 1976). It is controlled primarily by using clean or fungicide-treated seed. There is a demand for a reliable method for routine detection of the fungi in seeds. This method should preferably provide a quantitative measure of each of the fungi in a seed lot.

The only well-known and reliable method that discriminates between *P. teres* and *P. graminea* is a pathogenicity test on barley. *P. teres* produces disease symptoms on in-oculated barley leaves, whereas *P. graminea* cannot infect through the leaves, and leaf stripe disease only develops after seed infection. All other characteristics investigated, such as the morphology and the size of the conidia, the mycelial characteristics when grown *in vitro* (Smedegård-Petersen, 1971, 1977; Chidambaram *et al.*, 1973) and the ability to form a lilac pigment on Kietreiber's blotter (Knudsen, 1982) cannot clearly differ-entiate the fungi. Serological investigations have also shown that the fungi are very similar (Burns, 1992). Polyclonal antisera raised against soluble proteins was unspecific even after cross-absorption (Husted, 1993a). Acid phosphatase isoenzyme analysis is, however, able to differentiate between the three fungi but this method is not suitable for routine testing of barley seeds (Johansen *et al.*, 1993).

Our aim was to identify species-specific DNA fragments, which could be used in routine tests of barley seeds (Husted, 1993b). Preliminary results using RAPD showed a high degree of intraspecific variation, but one set of primers distinguished *P. graminea* and *P. teres* isolates (Reeves and Ball, 1991).

PROTOCOLS

Fungal DNA extraction and cloning – protocol

Materials

1. NIB-buffer: 0.1 M NaCl, 30 mM Tris-HCl, 10 mM EDTA, 10 mM β-mercaptoethanol, 0.5% (v/v) NP-40 (or Triton X-100), pH 8.0.
2. Homogenizing buffer: 0.1 M NaCl, 0.2 M sucrose, 10 mM EDTA.
3. Lysing buffer: 0.25 M EDTA, 0.5 M Tris-HCl, 2.5% SDS, pH 9.2.
4. TE-buffer: 10 mM Tris-HCl, 1 mM EDTA, pH 8.0.
5. 1% agarose gel (Sigma A-6877. Type II: Medium EEO) in TAE-buffer. TAE: 0.04 M Tris-acetate, 1 mM EDTA.
6. Restriction enzymes from Boehringer or Promega.
7. Table centrifuge: Biofuge A, Heraeus CHRIST.
8. Horizontal gel electrophoresis system: Model H4, BRL.

Method

Culturing of fungi

1. Flasks with Fries' medium (Smedegård-Petersen, 1976) enriched with 1% yeast extract are inoculated with spores and/or mycelial pieces taken from a fresh culture grown on grass agar.
2. Incubation for 3 days on a rotary shaker 110 rpm, 24°C.
3. Harvest the cultures by vacuum filtration and wash the mycelium with double distilled (dd) H_2O.
4. Lyophilize the mycelium and store it over silica gel at −20°C until use.

DNA extraction (modified from Panabières *et al.*, 1989)

1. Grind 100 mg lyophilized mycelium in a mortar and transfer it to an Eppendorf tube.
2. Suspend in 0.7 ml NIB-buffer.
3. Centrifuge at 13,000 rpm for 1 min.
4. Resuspend pellet in 0.7 ml NIB-buffer.
5. As step 3.
6. Resuspend pellet in 0.5 ml homogenizing buffer.
7. Add 0.125 ml lysing buffer.
8. Incubate at 55°C for 30 min.
9. Centrifuge 13,000 rpm for 15 min. Transfer supernatant to a new tube.
10. Add 10 µl RNase A (10 mg ml^{-1}) and incubate at 37°C for 30 min.
11. Extract twice with 1 vol phenol : chloroform : isoamylalcohol (25 : 24 : 1). Phase separation accomplished by 13,000 rpm for 15 min.

12. Extract twice with chloroform:isoamylalcohol (24:1), 13,000 rpm for 5 min.
13. Precipitate DNA using 1 vol. of cold 96% EtOH.
14. Centrifuge at once at 13,000 rpm for 1 min.
15. Wash pellet in a large amount of 70% EtOH (500–750 μl).
16. Air-dry pellet and resuspend in 50 μl TE-buffer.
17. Check concentration by running 2 μl on a 1% agarose gel in TAE-buffer together with digested λ-DNA of known concentration.
18. Check quality by digestion of about 500–750 ng of DNA, e.g. by *Eco*RI.

Cloning

1. *Eco*RI- or *Pst*I-digested total DNA is cloned into the plasmid vector pGem-3Z (Pheiffer and Zimmerman, 1983).
2. *Escherichia coli* DH5α is transformed and plated on to LB agar containing ampicillin and the chromogenic substrate, X-Gal (Hanahan, 1985).
3. White recombinant bacterial colonies are picked and grown in small-scale cultures.
4. Plasmid DNA is isolated using standard protocols (Maniatis *et al.*, 1982).

Screening for species-specific fragments – protocol

Materials

See below for RFLP and dot-blot analysis.

Method

1. Plasmid DNA from recombinant clones are isolated according to Maniatis *et al.* (1982).
2. 100 ng of plasmid DNA are dot blotted as described below for dot-blot analysis. DNA from plasmids containing large inserts (> 5 kb) is double digested with the restriction enzyme used for cloning and one more as a way to get more smaller fragments. This digest is run in a 1% agarose gel and transferred to Hybond-N as described below for RFLP analysis.
   In both cases two identical membranes are prepared.
3. Prehybridization as described below.
4. Hybridization conditions as described below, but washing twice in 0.2 × SSC and not in 0.1 × SSC. The two identical membranes are hybridized to: (i) labelled total DNA from the organism from which the fragments originate; (ii) labelled total DNA from the closely related organism, respectively.
5. Clones hybridizing to either *P. teres* or *P. graminea*, but not both, are selected and tested further.

RFLP analysis – protocol

Materials

1. Restriction enzymes, gels and electrophoresis buffer as above.
2. Denaturing solution: 1.5 M NaCl, 0.5 M NaOH.
   Neutralizing solution: 1.5 M NaCl, 0.5 M Tris-HCl, pH 7.2, 1 mM EDTA.
3. 20 × SSC: 3.0 M NaCl, 0.3 M Na$_3$-citrate, pH 7.0.

4. Prehybridization solution: $10 \times$ Denhardt, 50 mM sodium phosphate, pH 6.5, 0.1% Na-pyrophosphate, 0.75 M NaCl, 5 mM EDTA, 100 μg of denatured salmon sperm DNA added per ml.
   Hybridization solution: As prehybridization, but without salmon sperm DNA.
5. Hybridization oven: Hybaid.
6. Nylon membrane: Hybond-N from Amersham.
7. Probes labelled by [^{32}P]dCTP according to Feinberg and Vogelstein (1983). Many different probes have been selected and tested, but five of them more thoroughly tested. These are designated: (i) P.gr. P6–0.6, (ii) P.gr. E26–40, (iii) P.gr. E26–1.65, (iv)P.t.t.P65–3.2 and (v) P.t.m.E138–1.05. (The probes may be obtained from the author, K. Husted, Research Centre for Plant Protection, Danish Institute of Plant and Soil Science, Lottenborgvej 2, DK-2800 Lyngby, Denmark.)

Method

1. 750 ng fungal DNA are digested by appropriate restriction enzyme (most used here are *Eco*RI and *Pst*I).
2. Electrophoresis through 1% agarose gel.
3. Blotting to Hybond-N according to manufacturer's instructions.
4. Prehybridize at 68°C for at least 4 h in a hybridization oven.
5. Replace prehybridization solution with hybridization solution, add denatured probe and hybridize at 68°C for about 18 h in a hybridization oven.
6. Wash membrane: twice in $2 \times$ SSC containing 0.5% SDS at room temperature for 5 and 15 min; once in $0.2 \times$ SSC containing 0.5% SDS at 68°C for 1 h; and once in $0.1 \times$ SSC containing 0.5% SDS at 68°C for 1 h.
7. Expose membrane to X-ray film (XAR, Kodak) for 1–7 days at −80°C with an intensifying screen.

Dot-blot analysis – protocol

Materials

1. Dot-blot apparatus: Bio Rad.
2. Denaturing and neutralizing solution: as for RFLP analysis.
3. Hybridization solutions: as for RFLP analysis.

Method (manual application)

1. 50–100 ng of DNA is spotted on to membrane in approximately 2 μl aliquots. Allow to dry between the application of each aliquot.
2. Wet membrane in denaturing solution for 1 min.
3. Wet membrane in neutralizing solution for 1 min.
4. Fix DNA on a UV transilluminator for 3–4 min. Air dry.
5. Hybridization and washing as steps 4–7 for RFLP analysis.

Dot blotting using a dot-blot manifold apparatus (modified from Yao *et al.*, 1990)

1. DNA sample is made up to 0.4 M NaOH by adding 1 μl 4 M NaOH and H$_2$O up to 10 μl. (A larger volume might be used if the concentration of DNA samples is very low.)

2. Incubate for 10 min at room temperature.
3. Add 10 μl 2 M NH₄OAc (equal amount to the total volume from step 1).
4. Prewet Hybond-N membrane in ddH₂O for 2 min and in 20 × SSC for 30 min before use in dot-blot apparatus.
5. Dot blot 100 μl 1 M NH₄OAc to each spot.
6. Dot blot the DNA samples.
7. Wash in 100 μl 1 M NH₄OAc to each spot.
8. Wet in 4 × SSC for 5 min.
9. Fix DNA on a UV transilluminator for 3–4 min. Air dry.
10. Hybridization and washing as above.

INTERPRETATION AND EXAMPLES OF RESULTS

Extraction of total DNA yielded 50–250 μg DNA per gram of freeze dried mycelium. In a few cases the digestibility of the extract was low, probably because of a large amount of polysaccharides.

Plasmid DNA from more than 1000 recombinant clones of *P. graminea* and *P. teres* were isolated and screened. The high degree of similarity between the fungi was confirmed, in that only 12 fragments hybridized to total DNA from either *P. graminea* or *P. teres*, but not both. Subsequent testing by Southern blots and dot blots left five probes of interest: (i) P.gr. P6–0.6 specific for *P. graminea*; (ii) P.gr. E26–4.0 specific to *P. graminea* and *P. teres* f. *maculata*; (iii) P.gr. E26–1.65 (a subclone from P.gr. E26–4.0) specific to *P. graminea*; (iv) P.t.t.P65–3.2 specific to *P. teres* f. *teres* and giving weaker hybridization to *P. teres* f. *maculata*; and (v) P.t.m.E138–1.05 specific to *P. teres*.

In all cases, a weak cross-hybridization to a few isolates (< 10%) of the other species were found. In RFLP analyses this phenomenon had no practical importance, since all

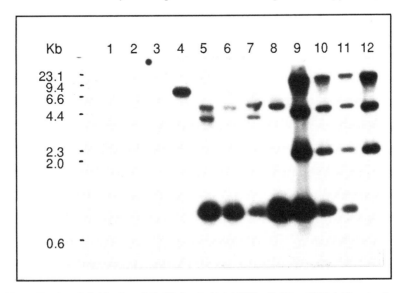

Fig 27.1. Southern hybridization of *Eco*RI-digested total DNA with P.t.m.E138–1.05 probe. Lanes 1–4: *P. graminea*; lanes 5–8: *P. teres* f. *teres*; lanes 9–12: *P. teres* f. *maculata*.

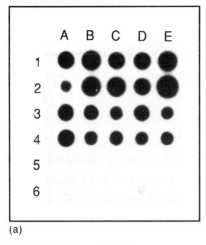
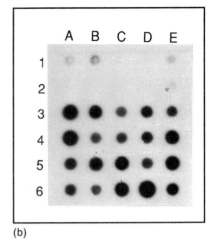

(a)   (b)

Fig 27.2.   Dot-blot analysis. Rows 1 and 2: *P. graminea* isolates; rows 3 and 4: *P. teres* f. *maculata* isolates; rows 5 and 6: *P. teres* f. *teres* isolates.
(a) Hybridized to P.gr.E26–4.0 probe. (b) Hybridized to P.t.m.E138–1.05 probe.

probes gave a clear differentiation of the three fungi based on the RFLP pattern. A degree of intraspecific variation appeared for all probes (e.g. in Fig. 27.1). In dot-blot analyses, the cross-hybridizations were in nearly all cases clearly distinguishable from the strong homologous hybridization. Comparison of results from more than one probe always gave a clear identification (e.g. Fig. 27.2).

For all probes no cross-hybridization was found to DNA from one isolate of the following species: *Pyrenophora avenae*, *Alternaria* spp., *Fusarium* spp., *Aspergillus* sp., *Microdochium nivale*, *Penicillium aurantiogriseum*, *Penicillium* cf. *hordei*, and *Acremonium* sp. respectively. The two probes P.gr.E26–4.0 and P.gr.E26–1.65 both gave a very weak cross-hybridization to two isolates of *Pyrenophora bromi* and one isolate of *Cochliobolus sativum*. Finally an isolate of *Epicoccum purpurascus* showed cross-hybridization to all five probes except the P.t.m.E138–1.05 probe.

REFERENCES

Burns, R. (1992) In: COST 88. Methods of early detection and identification of plant diseases. *Report of 1992 Activities.* BRIDGE, Commission of the European Communities.

Chidambaram, P., Mathur, S. B. and Neergaard, P. (1973) Identification of seed-borne *Drechslera* species. *Friesia* 10, 165–207.

Feinberg, A. and Vogelstein, B. (1983) A technique for radiolabelling DNA restriction endonuclease fragments to high specific activity. *Proceedings of the National Academy of Sciences of the United States of America,* 132, 6–13.

Hanahan, D. (1985) Studies on transformation of *Escherichia coli* with plasmids. *Journal of Molecular Biology* 166, 557–580.

Husted, K. (1993a) Differentiation of *Pyrenophora graminea* and *Pyrenophora teres*. II. Antiserum against soluble mycelial proteins. *Danish Journal of Plant and Soil Science* 96, 399–404.

Husted, K. (1993b) DNA probes for detection and differentiation of *Pyrenophora teres* and *P. graminea*. *Applied and Environmental Microbiology* (submitted).

Johansen, L. H., Husted, K., Olson, L. W. and Heide, M. (1993) Differentiation of *Pyrenophora graminea* and *Pyrenophora teres*. I. Gel electrophoresis and isozyme analysis of soluble mycelial proteins. *Danish Journal of Plant and Soil Science* 96, 391–398.

Knudsen, J. C. N. (1982) Production of pigments by *Pyrenophora graminea* and *Pyrenophora teres* with special reference to the use of this characteristic in seed disease testing. *Seed Science and Technology* 10, 357–363.

Maniatis, T., Fritsch, E. F. and Sambook, J. (1982) *Molecular Cloning: A Laboratory Manual*. Cold Spring Harbor Laboratory, Cold Spring Harbor, NY.

Panabières, F., Marais, A., Trentin, F., Bonnet, P. and Ricci, P. (1989) Repetitive DNA polymorphism analysis as a tool for identifying *Phytophthora* species. *Phytopathology* 79, 1105–1109.

Pheiffer, B. H. and Zimmerman, S. B. (1983) Polymer-stimulated ligation: Enhanced blunt- or cohesive-end ligation of DNA or deoxyribooligonucleotides by T4 DNA ligase in polymer solutions. *Nucleic Acids Research* 11, 7853–7871.

Reeves, J. and Ball, S. (1991) Preliminary results on the identification of *Pyrenophora* species using DNA polymorphisms amplified from arbitrary primers. *Plant Varieties and Seeds* 4, 185–189.

Smedegård-Petersen, V. (1971) *Pyrenophora teres* f. *maculata* f. nov. and *Pyrenophora teres* f. *teres* on barley in Denmark. *Royal Veterinary and Agricultural University Yearbook* 124–144.

Smedegård-Petersen, V. (1976) Pathogenesis and Genetics of Net-spot Blotch and Leaf Stripe of Barley Caused by *Pyrenophora teres* and *Pyrenophora graminea*. Thesis. Royal Veterinary and Agricultural University, Copenhagen. 176pp.

Smedegård-Petersen, V. (1977) Inheritance of genetic factors for symptoms and pathogenicity in hybrids of *Pyrenophora teres* and *Pyrenophora graminea*. *Phytopathologische Zeitschrift* 89, 193–202.

Yao, C. L., Frederiksen, R. A. and Magill, C. W. (1990) Seed transmission of sorghum downy mildew: Detection by DNA hybridization. *Seed Science and Technology* 18, 201–217.

28 Monoclonal Antibodies for the Detection of *Pyrenophora graminea*

R. BURNS, M. L. VERNON AND E. L. GEORGE

Scottish Agricultural Science Agency, East Craigs, Edinburgh EH12 8NJ, UK.

BACKGROUND

Pyrenophora graminea (Ito and Kurib. apud Ito) (anamorph *Drechslera graminea* (Rabenh. ex Schlecht.) Shoem. causes leaf stripe of barley. Infection in the seedlings can cause stunting and in severe cases lead to death. Characteristic symptoms are chlorotic or yellow stripes on leaves and sheaths. Leaf shedding usually occurs later as the necrotic leaf areas begin to dry out. Ears may not emerge properly or be discoloured and as a result the grain production of affected plants is severely reduced. The disease is found in most parts of the world where barley is grown. Transmission is by mycelium lying in the pericarp, production of perithecia is uncommon. Conidia are produced and account for floral infection and subsequent seed contamination (Ellis and Waller, 1973a). This disease used to be of limited importance as it was easily controlled with applications of mercury-based fungicides. In recent years mercury-resistant isolates have become more prevalent leading to larger losses. Barley seed is routinely tested for the presence of the organism by incubation on potato dextrose agar with visual inspection of the resultant colonies. *Pyrenophora teres* Drechsler (anamorph *Drechslera teres* (Sacc.) Shoem. causes net blotch disease of barley. Seed-borne inoculum produces the initial infection and is characterized by pale lesions on the primary leaves. These darken as the infection progresses and form spots and streaks. Secondary infection by conidia occurs leading to pale brown blotches on secondary leaves. These lesions have characteristic networks of darker lines within them (Ellis and Waller, 1973b).

It is important in seed certification to be able to differentiate the two species. Both, however, are very variable and some isolates are not easily distinguished by inspection of agar-grown colonies. Other methods of identification can be used but these can be very time consuming. An assay which could be used to rapidly differentiate the two species would be useful. It is also important to know the extent of contamination within seed lots and a rapid quantitative test could be of great value.

Nine monoclonal-antibody-secreting cell lines were produced to *P. graminea*. The

199

antibodies show variable reactions to isolates of the two species in immunofluoresence, ELISA and dot-blot.

Monoclonal antibody production

Polyclonal antibodies were produced by inoculating a New Zealand white rabbit with whole mycelium of *P. graminea*. Serum was assayed for specificity and found to react to both *Pyrenophora* species.

Monoclonal-antibody-secreting cell lines were produced according to traditional protocols, (Kennet *et al.*, 1978) using whole mycelium of *P. graminea* as the immunogen. Antibodies produced by this method showed no differentiation between the two species. Therefore, an approach was undertaken based on neonatal toleziation. Nine neonatal mice were immunized with mycelial washes of *P. teres*. Animals were immunized on the four consecutive days after birth to produce immunotolerance to antigens occurring in *P. teres* (Hsu *et al.*, 1990a,b) (Burns *et al.*, in preparation). At the age of 8 weeks the mice were immunized with mycelial washes of a mixture of isolates of *P. graminea*. Immunizations were repeated after 14 days and then 30 days. Cell fusions of splenocytes to NS-O myeloma cells were then performed. Screening of hybridoma supernatants was carried out by ELISA against mixtures of *P. graminea* and *P. teres* isolates.

Nine cell lines were isolated that appeared to discriminate between the two species. Further investigations into other isolates of the fungi showed that each antibody was specific to a number of isolates of *P. graminea* but not the full species range that was originally thought. In addition, some antibodies reacted to a number of *P. graminea* isolates and a few isolates of *P. teres*.

PROTOCOLS

Immunofluorescence (IF)

Materials

1. PBS.
2. Goat antimouse FITC conjugate.
3. Anti-quench mounting medium.

Method

1. Scrape mycelium from agar slope-cultures of the fungi and suspend in PBS.
2. Wash twice with PBS by centrifugation.
3. Incubate in MAb for 1 h.
4. Suspend mycelium in goat antimouse FITC conjugate and incubate for 1 h.
5. Wash three times with PBS by centrifugation.

6. Mount mycelium under coverslips in anti-quench medium.

7. View using UV microscope.

Dot-ELISA for detection of fungi in barley seed

Materials

1. Nitrocellulose (Schleicher and Schuell).

2. Blocking buffer – 10% powdered milk/5% BSA in TBS.

3. Substrate buffer – 20 ml *N,N*-dimethylformamide + 0.05 g naphthol AS-BI phosphate + 20 ml distilled H_2O.

4. Adjust pH to 8.0 with 0.1 M Na_2CO_3.

5. Add above to 600 ml distilled H_2O+360 ml Tris, pH 8.3.

6. Substrate – Fast Red TR salt (Sigma F8764).

Method

1. Incubate seed material in dH_2O for 24 h.

2. Freeze seed at −20°C for 24 h to destroy embryo.

3. Place seeds on nitrocellulose sheets in humid boxes and incubate 72 h.

4. Remove seed and block nitrocellulose.

5. Treat nitrocellulose with hybridoma supernatant.

6. Wash with TBS.

7. Incubate with antimouse alkaline phosphatase conjugate (Sigma A-0162).

8. Wash in TBS.

9. Incubate with 0.05 g Fast Red TR salt in 50 ml substrate buffer.

RESULTS

Immunofluorescence microscopy

Indirect immunofluorescent microscopy was also carried out on mycelia from isolates of the two species. Specificity to a number of isolates by each of the antibodies was once again observed. In most cases the outer surface of the hyphae fluoresced indicating a positive result. Fluorescence was, however, not even; various areas being more strongly visualized than others (Fig. 28.1). In one case when an Italian isolate of *P. graminea* was challenged with MAb 1, no fluorescence of the surface occurred; however, chlamydospores within the hyphae were strongly illuminated. Most cell lines produced antibodies which reacted to approximately 50% of isolates of both *P. graminea* and *P. teres* in IF. The isolates to which the antibodies reacted were not the same in all cases. One antibody reacted to 70% of *P. graminea* isolates tested and to only 20% of *P. teres*. This antibody was later used to develop a Dot-ELISA assay in seed material.

Dot-ELISA on nitrocellulose

It was apparent from results from DAS-ELISA that there was insufficient antigen expressed in seed material to be detected by this method. Various attempts at incubating seed material in nutrient broth, PBS and water failed to produce a signal in plate ELISA. Mechanical damage to the seed coat prior to these treatments did not improve on the results obtained.

An attempt was then made to develop a Dot-ELISA on nitrocellulose after allowing development of mycelial growth on to it from seed material. The results from this appear to be encouraging and a positive (red) patch appears on the nitrocellulose where a positive sample has been incubated. Seed lots of known levels of contamination have been assayed by this system and appropriate numbers of seeds giving positive reactions have been seen.

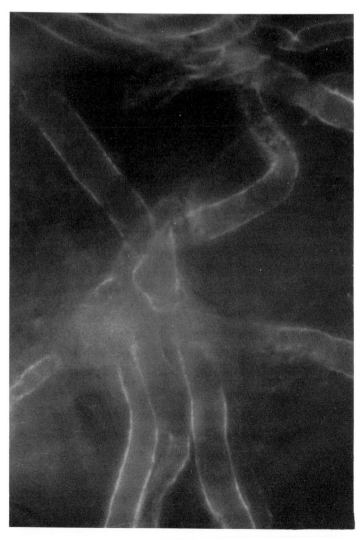

Fig 28.1. *Pyrenophora graminea* mycelia showing surface immunofluorescence with MAb 1.

PROSPECTS AND LIMITATIONS

The antibodies used in these assays are less than ideal as they are known to react to isolates of both species but an assay system for the amplification and detection of the fungi was developed.

Antibodies were challenged with a large number of other fungi by both the described methods and no cross-reactivity was seen. The antibodies appear to be genus specific with isolate variability.

The development of the Dot-ELISA assay on barley seed now gives us a system to work towards. The development of more MAbs to *Pyrenophora graminea* will soon be underway and it is hoped that more specific antibodies will be produced. The use of antibodies to visualize fungal structures by IF is interesting from a taxonomic viewpoint but would have little application in the routine screening of large seed lots for presence of disease. The large variability in this genus will probably be the largest problem, as isolates of one species can closely resemble the other. Their very different epidemiology, however, would suggest intrinsic differences which we may be able to exploit for the production of more MAbs.

REFERENCES

Ellis, M. B. and Waller, J. M. (1973a) *Pyrenophora graminea. Commonwealth Mycological Institute Descriptions of Pathogenic Fungi and Bacteria*, No. 388.

Ellis, M. B. and Waller, J. M. (1973b) *Pyrenophora teres. Commonwealth Mycological Institute Descriptions of Pathogenic Fungi and Bacteria*, No. 390.

Hsu, H. T., Lee, I., Davis, R. E. and Wang, Y. C. (1990a) Immunisation for generation of hybridoma antibodies specifically reacting with plants infected with a mycoplasma like organism (MLO) and their use in detection of MLO antigens. *Phytopathology* 80, 946–950.

Hsu, H. T., Wang, Y. C., Lawson, R. H., Wang, M. and Gonzalves, D. (1990b) Splenocytes of mice with induced immunological tolerance to plant antigens for construction of hybridoma secreting tomato spotted wilt virus-specific antibodies. *Phytopathology* 80, 158–162.

Kennet, R. H., Denis, K. A., Tung, A. S. and Klinman, N. R. (1978) Hybrid plasmacytoma production: Fusions with adult spleen cells, monoclonal spleen fragments, neonatal spleen cells and human spleen cells. *Current Topics in Microbiological Immunology* 81, 495.

29 PCR for the Detection of *Pyrenophora* Species, *Fusarium moniliforme, Stenocarpella maydis,* and the *Phomopsis/Diaporthe* Complex

E. J. A. BLAKEMORE,[1] D. S. JACCOUD FILHO[2] AND J. C. REEVES[1]

[1] *National Institute for Agricultural Botany, Huntingdon, Road, Cambridge CB3 OLE, UK;*
[2] *Holder of a PhD Scholarship from RHAE/ CNPq–Agronomy Department/ UEPG/ PR/ Brazil.*

BACKGROUND

Pyrenophora

Pyrenophora graminea Ito and Kurib. causes leaf stripe in barley. *Pyrenophora teres* Drechsler f. *teres* and f. *maculata* also affect barley causing net blotch and a spot form of net blotch respectively, and all are capable of producing significant yield losses. *P. graminea* is primarily seed-borne; *P. teres* can also be seed-borne but is mainly transmitted locally from plant debris. It is important in seed health testing that each of these pathogens can be accurately distinguished from each other but this is difficult using traditional agar plate or blotter tests. Definitive confirmation of a diagnosis may require pathogenicity testing on host plants.

Fusarium moniliforme and *Stenocarpella maydis*

Fusarium moniliforme Sheldon (teleomorph: *Gibberella fujikuroi* ([Saw.]Wollenw.) and *Stenocarpella maydis* (Berk.) Sutton are both economically important fungal seed-borne pathogens causing seedling blight, stalk and cob rot of maize. In addition to their agricultural importance they can cause serious animal and human diseases by producing harmful mycotoxins. Both of these pathogens have worldwide distribution and vary in their pathogenicity and cultural characteristics.

At present seed health tests are carried out by incubating maize seed on moist blotters or on agar plates. This is slow, insensitive and requires considerable expertise for accurate identification of the pathogen.

Phomopsis/Diaporthe complex

Soyabean stem canker caused by *Diaporthe phaseolorum* f.sp. *meridionalis* (anam. *Phomopsis phaseoli* f.sp. *meridionalis*), is a new, aggressive variant of the seed-borne fungal pathogen complex *Diaporthe phaseolorum* and is the most important problem in the soyabean crop in Brazil at present. It was first detected in the 1989/90 season (Yorinori, 1990) and has spread to important soyabean production regions in Brazil. Crop losses can reach 80% (Yorinori, 1990). Recent data (EMBRAPA/CNPSo) from the 1991/92 harvest show that in some regions of Parana estimated losses were about 705,000 sacks; more than US$ 4 million.

The level of seed infection by this fungus is usually less than 1% (Yorinori, 1992) but nevertheless seed is the most effective means of dissemination of the disease over long distances.

Owing to these low levels of seed infection and the similarity of different species of the pathogen complex (*Phomopsis sojae, Phomopsis* spp.) causing pod, stem blight and seed decay, traditional seed testing methods are slow, and often it is difficult to distinguish conclusively the damaging organisms from less virulent variants.

General

Sensitive detection and accurate identification methods are required for all the pathogens detailed above. Preliminary investigations using RAPDs for profiling and character-ization have shown the potential of this technique for these purposes. RAPDs may also provide DNA fragments for use as probes or as a basis for the design of primers for use in a specific PCR-based seed test. This report presents results from an initial screening of primers for use in detection and identification in seed of these barley, maize and soyabean pathogens using RAPD-PCR.

PROTOCOLS

Pyrenophora

Fungal isolates

Five isolates of *Pyrenophora graminea*, two isolates of *P. teres* f. *teres* and one isolate of *P. teres* f. *maculata* were used. Their identity was confirmed by pathogenicity testing.

Protocol for PCR amplification of genomic DNA

DNA preparation   Inoculate each isolate into 100 ml of potato dextrose broth from a 7-day agar plate culture and incubate at 25°C in the dark on an orbital shaker. Collect the mycelium by filtration and extract the DNA using the Raeder and Broda (1985) protocol.

Amplification conditions Perform RAPD-PCR in 50 μl volumes using genomic DNA from each of the fungal isolates as follows. The primer sequence is 5′-GGACGATTCG-3′.

1. For a single 50 μl PCR add the following components to a 0.5 ml sterile microcentrifuge tube:

   5 μl of 10×PCR buffer (10 mM Tris-HCl pH 8.3, 50 mM KCl, 4 mM $MgCl_2$, 0.01% BSA);

   1 μl of 10 mM dNTP solution (Pharmacia);

   5 μl of 2 μM primer solution;

   X μl of molecular grade water, where $X = 50 -$ sum of the volumes of all the other reaction components;

   up to 50 ng of genomic DNA;

   0.25 μl (1.25U) of *Taq* polymerase (Perkin Elmer Cetus).

2. Spin briefly (1 s) in a microcentrifuge to mix.

3. Overlay with 50 μl of sterile liquid paraffin to prevent evaporation during heating.

4. Place in thermal cycler (MJ Research). Amplify over 45 cycles of 1 min at 92°C, 1 min at 35°C, and 1 min at 72°C.

5. Refrigerate products at 4°C at the end of the PCR reaction prior to analysis on a 1% agarose gel and detection of the PCR products by staining with ethidium bromide.

Fusarium moniliforme

Fungal isolates

Fourteen isolates of *Fusarium* spp. were obtained from South Africa.

Protocol for PCR amplification of genomic DNA

DNA preparation Inoculate the *Fusarium* isolates on to potato dextrose agar and incubate for 5–7 days. Using a Gilson micropipette with a sterile yellow pipette tip, take a minute amount of fungal mycelium from the top of the inoculated agar plate. Place this mycelium into a 1.5 ml Eppendorf tube containing 10 μl of TE buffer pH 8.0 and mix gently. This mixture is heated for 10 min at 100°C and then put on ice for 5 min.

Amplification conditions Perform RAPD-PCR, as for *Pyrenophora*, in 50 μl volumes using 1 μl of the supernatant of the boiled fungal mycelium as a template. The primer sequence is 5′-CGGTTTGGTC-3′.

Stenocarpella maydis

Fungal isolates

Fifteen isolates were obtained from South Africa.

Protocol for PCR amplification of genomic DNA

DNA preparation Prepare DNA in the same way as for the *Fusarium* isolates (see above).

Amplification conditions Perform RAPD-PCR as for the *Fusarium* isolates given above. The primer sequence is 5′-CGGTTTGGTC-3′. Perform steps 1 to 5 as for *Fusarium*.

Phomopsis phaseoli f.sp. meridionalis

Fungal isolates

Isolates of *Phomopsis phaseoli* f.sp *meridionalis* from Brazil (EMBRAPA-CNPSo), *Macrophomina phaseolina* and *Phoma* sp. were used. Pathogenicity tests were done to confirm the identity of the isolates of *P. phaseoli* f.sp. *meridionalis*.

Protocol for PCR amplification of genomic DNA

DNA preparation   Prepare DNA as for the *Fusarium* isolates (see above). In addition take minute amounts of conidia directly from pycnidia of isolates of *P. phaseoli* f.sp. *meridionalis*.

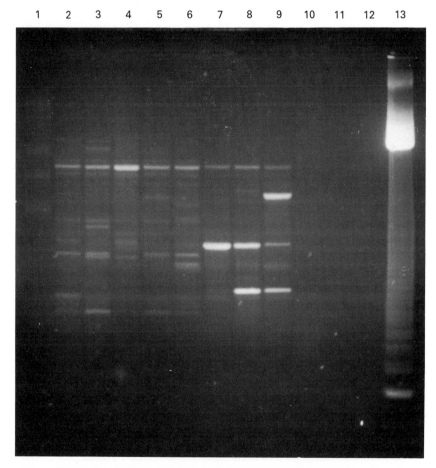

Fig 29.1.   DNA banding patterns amplified from all *Pyrenophora* isolates. Lanes 1 and 13, molecular weight markers; lanes 2 to 6, *Pyrenophora graminea*, lanes 7 and 8, *P. teres* f. *teres*, lane 9, *P. teres* f. *maculata*, lanes 10 to 12, control reactions.

1   2   3   4   5   6   7   8   9   10   11 12 13 14 15 16 17

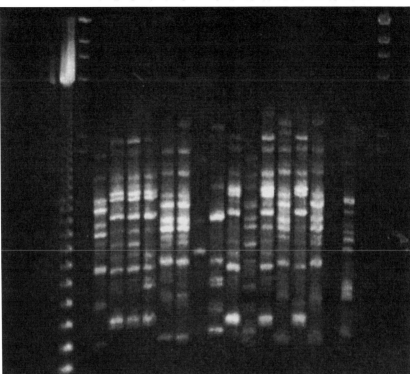

Fig 29.2.   DNA banding patterns of amplified DNA from *Fusarium* spp. isolated from maize seed. All the *Fusarium moniliforme* isolates show a distinctive DNA banding pattern (lanes 2,3,4,9,11 and 13) which differs from the other *Fusarium* spp. samples on this gel. Lanes 15,16 and 17 are controls.

Amplification conditions   Perform RAPD-PCR as for the *Fusarium* isolates above but 25 μl or 50 μl reaction volumes can be used with appropriate adjustment. The following primers are used:

primer 4:5′-ACTCAGCCAC-3′
primer 8:5′-GCAGAGCTAA-3′
primer 70:5′-CGTAGTGGTG-3′
Steps 1 to 5 as for *Fusarium*.

RESULTS AND DISCUSSION

Pyrenophora

Figure 29.1 shows that this primer produces a pattern of major bands which reflect the similarities between the isolates but which also reveals differences which correlate with their pathogenicity grouping and which could be used for differentiation.

All isolates give a relatively intense high molecular weight band but in *Pyrenophora graminea* no other equally intense bands are present. In *Pyrenophora teres* f. *teres* there is

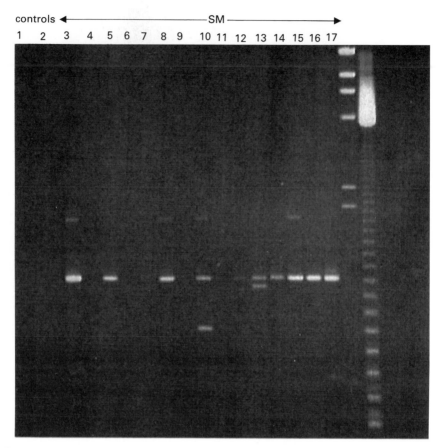

Fig 29.3. DNA banding patterns of amplified DNA from different isolates of *Stenocarpella maydis* (SM) obtained from South African maize seed.

one lower molecular weight band common to both isolates with one isolate having an additional lower molecular weight band. Both these bands are present in *Pyrenophora teres* f. *maculata* with a further higher molecular weight band.

Such DNA profiles could be used to supplement other phenotypic data on isolates obtained from standard seed tests and this would provide an immediate if somewhat ungainly use for the technique. A more sophisticated approach would be to excise a band or bands to use as probes for detection and identification.

Other primers (Reeves and Ball, 1991) reveal polymorphisms within this complex of pathogens which are not so clearly correlated with disease symptoms. Using such primers with a larger number of isolates from a range of sources, banding patterns can be generated, scored and submitted to an appropriate cluster analysis to investigate the relationships between isolates.

Fusarium moniliforme

Figure 29.2 shows that this primer produces a distinctive DNA banding pattern for *F. moniliforme* which differs from the other *Fusarium* spp. on this agarose gel. With a

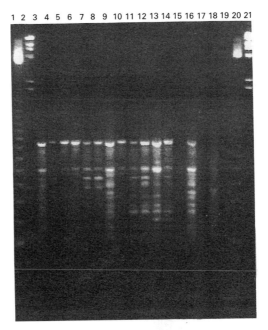

Fig 29.4. DNA banding patterns from seven different isolates of *Phomopsis phaseoli* f.sp. *meridionalis* amplified using primer 70. Lanes 3,5,7,9,11,13 and 16 are from conidial templates. Lanes 4,6,8,10,12,14 and 15 are from mycelial templates. Lanes 17,18 and 19 are control reactions. Lanes 1,2,20 and 21 are molecular weight markers.

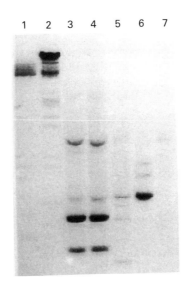

Fig 29.5. DNA banding patterns obtained using primer 4 from two different isolates of *Phomopsis phaseoli* f.sp. *meridionalis* (lanes 3 and 4); one isolate of *Macrophomina phaseolina* (lane 5) and an isolate of *Phoma* sp. (lane 6). Lanes 1 and 2 are molecular weight markers and lane 7 is a control reaction.

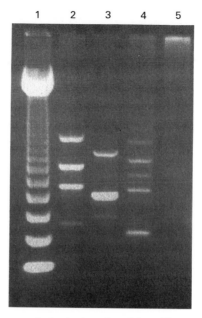

Fig 29.6. DNA banding patterns obtained using primer 8 from *Phomopsis phaseoli* f.sp. *meridionalis* (lane 2); *Macrophomina phaseolina* (lane 3) and *Phoma* sp. (lane 4). Lane 1 is a molecular weight marker and lane 5 is a control reaction.

widespread pathogen like *Fusarium moniliforme* it will be necessary to establish that this banding pattern is constant or retains constant discriminatory features at a species level irrespective of any geographical variation.

Stenocarpella maydis

Figure 29.3 shows that this primer produces one major band which is in common with all the isolates where there was amplification. Although this primer shows little poly-morphism it may be that the single main band could be used as a probe, as described above.

Phomopsis phaseoli f.sp. *meridionalis*

Figure 29.4 shows seven different isolates of *P. phaseoli* f.sp. *meridionalis* amplified using primer 70 from mycelial and conidial templates. There were no major differences between banding patterns derived from either source or between isolates. No differences between the banding patterns obtained from mycelial and conidial templates were expected. These patterns will be compared with those obtained from extracted DNA. Although superficially trivial, this comparison of banding patterns from mycelium, conidia and extracted DNA will confirm the validity of amplification directly from material taken indiscriminately from pure culture on an agar plate.

Figure 29.5 shows amplification using primer 4 of two different isolates of *P. phaseoli* f.sp. *meridionalis*; one isolate of *M. phaseolina* and an isolate of *Phoma* sp. There are no differences between the isolates of *P. phaseoli* f.sp. *meridionalis* but there were clear differences between these and the other species represented.

Figure 29.6 shows strong amplification of *P. phaseoli* f.sp. *meridionalis*, *M. phaseolina* and *Phoma* sp. using primer 8. Each of the single isolates produces a distinct pattern based on the major bands.

LIMITATIONS

The work reported here is at an early stage and requires further development before these techniques can be incorporated into the design of new tests for the detection in seed of the pathogens investigated. Such tests could be based on the profiles generated using RAPDs, which would aid identification where this is a primary difficulty. The ability to amplify from crude mycelial preparations is an important factor. Alternatively, DNA probes can be obtained by excising bands from the gel that are species specific (Blakemore *et al.*, 1992). Specific primers, obtained by sequencing such a band, could be used to develop a PCR-based test. This could provide the means for both accurate identification and sensitive, rapid detection. Before this can be done for the pathogens mentioned here, more screening of isolates of the pathogens and of other co-isolated organisms is required. This work is currently underway.

REFERENCES

Blakemore E. J. A., Reeves J. C. and Ball S. F. L. (1992) Polymerase chain reaction used in the development of a probe to identify *Erwinia stewartii*, a bacterial pathogen of maize. *Seed Science and Technology* 20, 331–335.

Raeder, U. and Broda, P. (1985) Rapid preparation of DNA from filamentous fungi. *Letters in Applied Microbiology* 1, 17–20.

Reeves, J. C. and Ball, S. F. L. (1991) Research note: Preliminary results on the identification of *Pyrenophora* species using DNA polymorphisms amplified from arbitrary primers. *Plant Varieties and Seeds* 4, 185–189.

Yorinori, J. T. (1990) Cancro da Haste da Soja. Embrapa-CNPSo, *Comunicado Tecnico* No. 44, Maio /90, 1–8.

Yorinori, J. T. (1992) Identification of *Diaporthe phaseolorum* f.sp. Soybean by the Blotter Test, In: *Proceedings of 23rd ISTA Congress*, Buenos Aires, Argentina, 2–4 November 1992, p. 107.

30 PCR for Detection of the Fungi that Cause Sigatoka Leaf Spots of Banana and Plantain

A. JOHANSON

Natural Resources Institute, Chatham Maritime, Chatham, Kent ME4 4TB, UK.

BACKGROUND

There are two forms of Sigatoka leaf spots which affect bananas – yellow Sigatoka, caused by the fungus *Mycosphaerella musicola* (Zimm.) Deighton (anamorph *Pseudocercospora musae*), and black Sigatoka, caused by *M. fijiensis* (Morelet) Deighton (anamorph *Paracercospora fijiensis*) (Stover, 1980).

Both leaf spot diseases are of economic importance, but black Sigatoka develops much more rapidly, causes more severe defoliation, and is more difficult to control than yellow Sigatoka. In addition, black Sigatoka is pathogenic to plantain which under most circumstances is resistant to yellow Sigatoka. Yellow Sigatoka is present almost everywhere bananas are grown, but black Sigatoka is still absent from most of the Caribbean islands (with the exception of Cuba) and several Latin American and African countries.

Defoliation due to both diseases can cause both yield loss and a reduction in quality. If leaf spot is severe the fruit age physiologically causing premature ripening of the fruit in transit, and a condition known as 'yellow pulp'. In all commercial banana growing areas fungicides are used to control the diseases.

Although black Sigatoka can often be recognized visually, unambiguous diagnosis is complicated by the presence of other pathogens found on banana leaves. If yellow Sigatoka infection is heavy it can be almost impossible to distinguish from black Sigatoka by symptoms alone. Symptoms also vary depending on the cultivar and whether the plants have received oil and/or fungicide treatment. Isolation of the pathogen, which is most successfully achieved by ascospore discharge from necrotic leaf material, is often confounded by the absence of mature perithecia, and even when obtained in culture, *M. fijiensis* and *M. musicola* are not readily differentiated (Pons, 1990). A technique for rapid and accurate identification of these species therefore would be of great value in epidemiological research, particularly for disease-monitoring programmes in areas which have so far been free of the black Sigatoka pathogen.

Detection by amplification of ribosomal DNA

The approach developed for the detection of the *Mycosphaerella* species was the amplification of ribosomal DNA genes. Amplification of ribosomal genes is used for the genetic analysis of many organisms because of their ease of isolation and relatively high gene copy number. Although the nucleotide sequences of mature RNAs are highly conserved, both non-transcribed and transcribed spacer sequences, which often make up approximately half of the rRNA repeating unit, are usually poorly conserved (Nazar *et al.*, 1991). An area of the ribosomal DNA known as the ITS (internal transcribed spacer) region was sequenced from isolates of each of the two major species, and several other *Mycosphaerella* spp. found on banana leaves. This was done by amplifying the ITS1 region (between the 18S and 5.8S rDNA subunits) by PCR using standard primers (ITS1 and ITS2) (White *et al.*, 1990). Sequences were found in this region which were different for *M. fijiensis* and *M. musicola*. Two 21-oligonucleotide primers were constructed from these sequences: primer MF137 from the sequence of *M. fijiensis*; and primer MM137 from the sequence of *M. musicola*. An additional primer designed from the sequence of *M. musae* was also constructed (MUS781). These are used in association with primers situated within the large (25S) subunit of rDNA to produce specific products when total genomic fungal DNA is amplified by PCR. The location of these primers on the fungal ribosomal DNA is shown in Fig. 30.1.

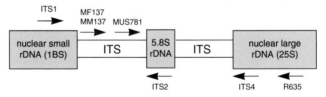

Fig 30.1. Location of PCR primers on ribosomal DNA.

DNA extracted from isolates of *M. fijiensis*, *M. musicola* and *M. musae* and amplified using primers MF137, MM137 and MUS781 respectively, in conjunction with primer R635, produces a single amplification product of approximately 1000 bp. If primer ITS4 is used instead of R635, single products of approximately 450 bp are observed. Each primer will amplify DNA only from the species from which its sequence was obtained (Fig. 30.2).

These single products specific for *M. fijiensis* and *M. musicola* can also be observed when DNA extracted from leaves of infected plants is amplified using these primers. Southern blotting has confirmed that these PCR products are of fungal origin, and not products of amplification of the plant DNA.

PCR can be carried out on DNA extracted from single-spored fungal cultures, or on total DNA extracted from infected leaf material. Preliminary results have also shown that single ascospores, discharged from infected leaves on to water agar, can serve as templates for the PCR reaction without pretreatment.

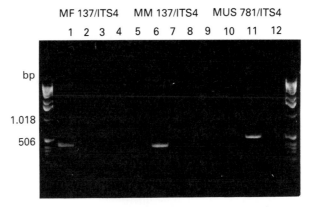

Fig 30.2. Ethidium bromide-stained agarose gel of amplification products of *Mycosphaerella* species DNA with ribosomal primers. Lanes 1–4, primers MF137/ITS4; lanes 5–8, primers MM137/ITS4; lanes 9–12, primers MUS781/ITS4. Lanes 1,5,9, *M. fijiensis*; lanes 2,6,10, *M. musicola*; lanes 3,7,11, *M. musae*; lanes 4,8,12, *M. minima*. DNA size marker is Gibco-BRL KB ladder.

PROTOCOL

Fungal isolates and genomic DNA extraction

1. Fungal cultures are grown by inoculating 50 ml Czapek Dox (Oxoid) liquid medium plus yeast extract (1 gl^{-1}) in a 250 ml conical flask with mycelial fragments of the isolates from cultures on PDA (Oxoid). Cultures are incubated on an orbital shaker (26°C) for 14 to 28 days depending on the growth of the isolate.

2. Mycelium is filtered through a Buchner funnel on to Whatman no. 3 filter paper and freeze-dried. Freeze-dried mycelium is ground in liquid nitrogen and total genomic DNA extracted using a protocol similar to those of Raeder and Broda (1985) and Lee *et al.* (1990) as outlined below.

3. Approximately 50 mg of ground mycelium is placed in a 1.5 ml microcentrifuge tube and 750 μl of extraction buffer (200 mM Tris-HCl pH 8.5, 250 mM NaCl, 25 mM EDTA, 0.5% SDS) is added. The tube is heated to 65°C for 1 h and 600 μl of phenol: chloroform added. This is then centrifuged for 30 min at 13,000 rpm (Hereaus Biofuge 15) and the supernatant removed to a new tube. Chloroform : isoamyl alcohol (24:1, 600 μl) is then added and the tube centrifuged at 13,000 rpm for another 10 min. The supernatant is then removed and DNA precipitated by the addition of 20 μl 3 M sodium acetate and 700 μl ice-cold absolute ethanol. The tube is then centrifuged briefly (13,000 rpm for 5 min) to pellet the DNA. Pellets are washed in 70% ethanol and then vacuum-dried. The DNA is resuspended in 100 μl sterile distilled water. RNase treatment is carried out either at this stage followed by a further chloroform extraction, or between the phenol and chloroform steps above.

Plant genomic DNA extraction

1. Green leaf samples are freeze-dried and ground in liquid nitrogen. Necrotic leaf samples are ground in liquid nitrogen without further drying.

2. Total plant DNA is extracted as described above for the fungal material.
3. After resuspension in 100 μl sterile distilled water, the DNA extract (usually darkly-pigmented at this stage) is then passed through a Magic DNA Clean-Up Column™ (Promega), according to the manufacturer's instructions, to remove phenolic compounds and pigments which inhibit subsequent DNA amplification.

Single spore isolation

1. Necrotic leaves with mature perithecia of the fungi are cut into pieces approximately 2 × 2 cm. These pieces are stapled to a 9 cm diameter filter paper in the lid of a Petri dish and soaked in distilled water for 15 min.
2. Excess water is then poured off the filter paper, and the Petri dish lid is placed on top of a Petri plate containing 1.5% water agar. The ascospores are allowed to discharge from the leaf on to the agar.
3. The plate is observed under a binocular microscope every 15 min to locate spores. Individual spores can be removed from the plate with as little agar as possible using a fine needle, and placed directly into a PCR tube. The PCR reagents are then added and the reaction carried out as described below.

PCR amplification and product analysis

1. Amplification reactions are performed in 50 mM KCl, 1.5 mM MgCl$_2$, 10 mM Tris-HCl pH 8.3, containing 200 μM of each dATP, dCTP, dGTP and dTTP (Pharmacia), 200 nM primer (Operon Technologies, Inc.), 1 unit of *Taq* DNA polymerase (Boehringer Mannheim) and approximately 0.5 μl of the genomic template DNA extract in a final volume of 50 μl.
2. DNA amplification by a two-step modification of the polymerase chain reaction is then performed in a thermal cycler (Techne PHC-3 or Genetic Research Instrumentation Mini-Cycler) programmed as follows: 94°C – 3 min; then 35 cycles of: 94°C – 1 min; 68°C – 2 min; followed by 1 cycle of: 72°C – 5 min.
3. PCR reaction samples are held at 4°C until analysed.
4. 10–20 μl of sample is then mixed with 4 μl sample loading buffer and run in a 1.5% agarose gel containing 0.5 μg ml^{-1} ethidium bromide at 100 V for up to 1 h with Tris-borate-EDTA as a running buffer.

Oligonucleotide primers

The sequence of primer R635, situated in the 25S subunit of rDNA was taken from Liu *et al.* (1991). Primer sequences were as follows: MF137, 5′ GGCGCCCCCGGAGGCCGTCTA 3′; MM137, 5′ GCGGCCCCCGGAGGTCTCCTT 3′; MUS781, 5′ TCTGAGTGAGGGCCTCCGGGT 3′; R635, 5′ GGTCCGTGTTTCAAGACGG 3′; ITS4, 5′ TCCTCCGCTTATTGATATGC. Primers were made by Operon Technologies, Inc., Alameda, CA, USA.

RESULTS

Sensitivity

The sensitivity of detection of target DNA by this method is very high. Target sequences can be detected in DNA extracts which have been prepared from 50 mg of dried fungal mycelium diluted up to 1 in 20,000 (Fig. 30.3). Results suggest that the use of primer ITS4, instead of R635, increases the sensitivity of detection and decreases the levels of smearing of DNA occasionally found when products are run on agarose gels. This is thought to be due to the greater errors which are likely in the amplification of a larger PCR product. A single ascospore picked up from a water agar plate can be detected by this method. The optimal annealing temperatures (T_m) of the 21-oligonucleotide primers are high, and advantage has been taken of this by reducing the PCR to a two-step process in which the annealing and extension cycles are combined into one step at 68°C. This allows amplification to be carried out more rapidly. Primer ITS4, however, has a T_m of only 58°C which may result in decreased sensitivity if annealing temperatures as high as 68°C are used. Further sequence information of this area of the 25S subunit could be obtained in order to design a similar primer with a higher T_m.

Specificity

Several other fungal species which are occasionally found saprophytically on banana leaves, such as *Aspergillus* sp., *Botryodiplodia theobromae* and *Fusarium moniliforme*, have been tested using these primers to check for any 'cross-reaction'. With the two-step PCR process, non-specific amplification is minimized, although occasionally very faint bands are observed for *Mycosphaerella minima* with primer MF137. Of all other fungal species tested, only two, *Colletotrichum gloeosporioides* and *C. musae* occasionally produce a single product with primer MM137. This product is at least 100 bp smaller than that produced from *M. musicola* and would not easily be confused with this product. Using this method, these fungi can be detected in leaves both at early stages of infection when leaves are still

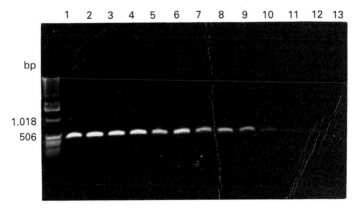

Fig 30.3. Ethidium bromide-stained agarose gel showing sensitivity of the ribosomal DNA primers (MF137/ITS4) to detect *Mycosphaerella fijiensis* DNA. Lanes 1–12 are serial dilutions of DNA from 50 mg freeze-dried fungal mycelium. Lane 1, 1 : 10; lane 2, 1 : 20; lane 3, 1 : 40 . . . lane 12, 1 : 20,000, lane 13, no DNA control. DNA size marker is Gibco-BRL KB ladder.

green and lesions are distinct, and at a much later stage when the leaves are dried and completely necrotic. Fungal DNA has been amplified from dried leaf samples up to 3 years old, suggesting that this technique may be useful for even older herbarium specimens.

LIMITATIONS

The most important limitation of the method is the preparation of the plant material for testing. At present it is usually necessary to purify the sample after DNA extraction, by passage through some kind of matrix which will remove substances inhibitory to the PCR reaction. Magic DNA Clean-Up columns (Promega) are used routinely for this purpose, but others, such as Qiagen tips have also been used successfully. Simple home-made Sephadex G200 spin columns or additional phenol:chloroform steps can also be effective. Older, dried samples contain more phenolics and are the most difficult to purify by these methods. It is occasionally necessary to pass samples twice through the purification process in order to remove all the inhibition and it is therefore recommended that adequate controls be used when testing samples. In particular, it is important to carry out a control PCR reaction in which the extract is 'spiked' with a known amount of target fungal DNA to identify possible false negatives which could be due to inhibition.

Simpler methods for extraction of plant DNA have been reported to have been successful with other plant species (Edwards *et al.*, 1991). These have not so far been useful in our experience with banana leaf tissue, as they do not remove the pigmented phenolics which inhibit the PCR reaction.

Work is continuing on the sample preparation aspect of the detection method. This is not, however, a problem if mature ascospores are present on the leaf sample, as these can then be discharged on to agar, picked up and placed directly in the PCR reaction. This does, however, require a good microscope, a knowledge of spore morphology and a steady hand. Since, however, the number of ascospores does not appear to be crucial to this method, another way of approaching this is to allow the spores to discharge on to a membrane, pieces of which could be cut out (with a hole punch) and placed directly into the PCR tube. No inhibitory effect has been observed with more than one ascospore, so the number picked is not important. Preliminary work suggests that this may be successful, but the choice of membrane type is crucial, as has been reported by other researchers (Bej *et al.*, 1991).

REFERENCES

Bej, A. K., Mahbubani, M. H., Dicesare, J. L. and Atlas, R. M. (1991) Polymerase chain reaction-gene probe detection of microorganisms by using filter-concentrates samples. *Applied and Environmental Microbiology* 57(12), 3529–3534.

Edwards, K., Johnstone, C. and Thompson, C. (1991) A simple and rapid method for the preparation of plant genomic DNA for PCR analysis. *Nucleic Acids Research* 19(6), 1349.

Lee, S. B., Milgroom, M. G. and Taylor, J. W. (1990) A rapid, high-yield mini-prep method for isolation of total genomic DNA from fungi. *Fungal Genetics Newsletter* 35, 23–24.

Liu, Z., Stewart, E. L. and Szabo, L. J. (1991) Phylogenetic relationships among *Cercospora* and allied genera on banana based on rDNA sequence comparisons. *Phytopathology* 81(10), 1240 (abstract).

Nazar, R. N., Hu, X., Schmidt, J., Culham, D. and Robb, J. (1991) Potential use of PCR-amplified ribosomal intergenic sequences in the detection and differentiation of Verticillium wilt pathogens. *Physiological and Molecular Plant Pathology* 39, 1–11.

Pons, N. (1990) Taxonomy of *Cercospora* and related genera. In: Fullerton, R. A. and Stover, R. H. (eds) *Sigatoka Leaf Spot Diseases of Bananas, Proceedings of an International Workshop held at San Jose, Costa Rica*, 28 March–1 April 1989.

Raeder, U. and Broda, P. (1985) Rapid preparation of DNA from filamentous fungi. *Letters in Applied Microbiology* 1, 17–20.

Stover, R. H. (1980) Sigatoka leaf spot of bananas and plantains. *Plant Disease*, 64, 750–755.

White, T. J., Bruns, T., Lee, S. and Taylor, J. W. (1990) Amplification and direct sequencing of fungal ribosomal RNA genes for phylogenetics. In: Innis, M. A., Gelfand, D. H., Sninsky, J. J. and White, T. J. (eds) *PCR Protocols: A Guide to Methods and Applications*. Academic Press, San Diego, pp. 315–322.

31 DNA Fingerprinting of the Potato Late Blight Fungus, *Phytophthora infestans*

A. DRENTH AND F. GOVERS

Department of Phytopathology, Wageningen Agricultural University, Binnenhaven 9, 6709 PD Wageningen, The Netherlands.

BACKGROUND

Phytophthora infestans (Mont.) de Bary is the causal agent of potato late blight which is considered to be the most damaging disease on potato worldwide (Hooker, 1981). *P. infestans* belongs to the oomycetes and is heterothallic with two known mating types, designated A1 and A2. Until recently the A2 mating type was confined to central Mexico whereas the A1 mating type was found all over the world. In the early 1980s, the A2 mating type was discovered in Europe (Hohl and Iselin, 1984) and subsequently in many countries throughout the world (Spielman *et al.*, 1991). With both mating types present in the field there is the possibility of sexual reproduction. Interaction between isolates of opposite mating type results in the formation of oospores, structures which may enable the fungus to survive in the soil during winter. Therefore, oospores may be a new potential inoculum source in The Netherlands contributing in early spring to the spread of the disease.

In the past decade, control of late blight in The Netherlands was troublesome because epidemics started earlier and were harder to control. Besides potatoes, tomatoes were attacked more severely than before. With regard to virulence, genetic diversity increased significantly which resulted in many new and more complex physiological races (Schöber, 1983; Rullich and Schöber, 1988; A. Drenth, unpublished). New allozyme genotypes were identified in western Europe indicating that not only the A2 mating type was introduced but that a complete new population consisting of A1 and A2 mating types had replaced the old A1 population in Europe (Spielman *et al.*, 1991). Failures to control late blight prompted us to investigate whether sexual reproduction, via the formation and survival of oospores, plays a role in the epidemiology of *P. infestans*.

According to Maynard Smith (1971), the effect of sexual reproduction can be twofold: (i) accelerated evolutionary adaptation of the population; (ii) more variability in the population. Accelerated evolutionary adaptation of the population is very difficult to measure. Increase of genetic variation can be measured directly provided that sufficient markers are available, either many different genetic markers or highly diagnostic markers such as DNA fingerprint probes and RAPDs.

Here we will first discuss the different markers which are currently available for *P. infestans*. Subsequently we will describe the probe and the methodology we routinely use for DNA fingerprinting of *P. infestans*. Finally, we present some examples of population genetic analyses of *P. infestans* using DNA fingerprinting.

Markers available for *P. infestans*

A prerequisite for population genetic analyses is the availability of genetic markers. The markers available for *P. infestans* can be divided into three categories namely, biologically significant markers, cytoplasmic markers and neutral markers.

Biologically significant markers

Biologically significant markers comprise mating type, virulence and fungicide resistance. The two mating types of *P. infestans*, A1 and A2, are readily detectable using tester isolates. Mating type is most likely governed by a single diallelic locus (Shaw, 1983). Virulence can be determined by assaying compatibility and incompatibility in a differential set of potato lines which contain the eleven known resistance genes (R-genes). Supposedly virulence/avirulence in *P. infestans* and resistance/susceptibility in potato act in a gene-for-gene matter as was demonstrated for a few genes (Spielman *et al.*, 1989, 1990). A problem for using virulence as an unambiguous marker is the lack of a universal differential set of potato lines and a universal virulence assessment method. This seriously limits the interpretation of data in the numerous papers concerning variation for virulence. Moreover, the stability of virulence phenotypes has been subject of much debate. Population genetic analysis to assess variation in *P. infestans* populations present before the 1980s was mainly based on virulence/avirulence phenotypes. The advantages of biologically significant markers is that they can be scored easily without the necessity to bring the fungus in pure culture. This allows large numbers to be screened at relatively low cost. There is no known selection on mating type. Virulence and fungicide resistance are traits on which there can be an enormous selection pressure exerted through the use of resistant potato cultivars or the application of fungicides. Strong selection on particular characteristics can dramatically alter their frequency in the fungal population, thus requiring careful interpretation of data. Besides a poor understanding of the genetics of these markers, the diagnostic value is limited because they are simply not numerous enough to identify fungal isolates unambiguously.

Cytoplasmic markers

The main characteristic of cytoplasmic markers is their extrachromosomal inheritance leading to intact transmission from parent to progeny. Markers in this category are mitochondrial DNA (mt-DNA) (Carter *et al.*, 1990; Goodwin, 1991) and double-stranded RNA (ds-RNA) (Newhouse *et al.*, 1992). The origin and the function of ds-RNA, which is found in only a limited number of isolates, is not known. Cytoplasmic markers are very useful to determine the global movement of plant pathogens, to infer phylogenetic relationships among different *Phytophthora* species and to follow anastomosis and cytoplasmic inheritance. However, they are not well suited for unambiguous identification of different isolates. The number of polymorphisms for mt-DNA and ds-RNA in *P. infestans* is rather limited. Because they are transmitted without recombination from parent to progeny they cannot reflect the true diversity in a sexual population.

Neutral markers

This category of markers is called neutral because the markers are not related to any biologically important characteristic. They are independent from each other and they are numerous. Neutral markers can be subdivided into allozyme markers and polymorphic DNA markers.

Enzymes which differ in electrophoretic mobility as a result of allelic diversity in a single gene are called allozymes. Among 50 different enzyme systems tested as neutral marker for *P. infestans*, only glucose-phosphate isomerase, peptidase and mannose-phosphate isomerase are polymorphic and resolve well on a starch gel (Tooley *et al.*, 1985; Spielman *et al.*, 1990).

A number of different techniques are currently being used to detect DNA polymorphisms. DNA polymorphisms result from differences in a DNA sequence due to substitutions, deletions, inversions or translocations. Moderately repetitive DNA probes or short oligonucleotide repeats allow simultaneous detection of restriction fragment length polymorphism (RFLP) at numerous loci in the genome. An example of a moderately repetitive probe that we use for DNA fingerprinting of *P. infestans* is described below.

Recently we demonstrated polymorphisms in a few *Phytophthora* species using $(GATA)_5$, $(GAG)_5$, $(GACA)_4$ and $(GTG)_5$ as probes (unpublished). An advantage of this so-called oligonucleotide fingerprinting technique is that the Southern blotting can be omitted by hybridizing the short oligonucleotides directly to size fractionated genomic DNA which is immobilized in dried agarose gels (Zischler *et al.*, 1989). DNA polymorphisms can also be detected by using the RAPD technique (Williams *et al.*, 1990). We used this technique for analysing DNA polymorphisms in different *P. infestans* isolates. Depending on the primer used, four to twelve fragments of *P. infestans* were amplified each being a specific, reproducible marker for a particular source of template DNA (unpublished). For identification of different isolates more than one RAPD primer is needed to get enough polymorphic bands.

In conclusion, it is evident that from the markers which are available for *P. infestans*, polymorphic DNA markers are the most suitable to conduct population genetic analyses. In particular, DNA fingerprinting, which enables detection of multiple loci, is a powerful method. A prerequisite for a fingerprint probe is that it detects loci which are dispersed all over the genome and are inherited independently.

DNA fingerprint probe RG-57

For *P. infestans*, moderately repetitive nuclear DNA fragments have been isolated which are very useful as probes for population genetic studies (Goodwin *et al.*, 1992a). One moderately repetitive DNA probe, designated RG-57, hybridizes to 26 different genomic *Eco*RI fragments, 23 of which have been shown to be polymorphic (Fig. 31.1). The banding pattern is stable through asexual reproduction and each hybridizing fragment appears to represent a unique genetic locus. Probe RG-57 allows virtually unambiguous identification of different isolates. Through genetic analysis of three different crosses in which 17 bands segregated it was shown that none of these bands is allelic. Only two pairs of co-segregating loci were identified. Each locus can be identified by the presence or absence of a hybridizing fragment (Goodwin *et al.*, 1992a). All *Eco*RI fragments hybridizing to probe RG-57 are between 1.1 and 18 kb and can be easily separated from each other on a 0.8% agarose gel. Use of two allele systems, particular those of the presence/absence type, avoid ambiguities in typing (Weir, 1992). This is in contrast to other DNA fingerprinting patterns such as variable number of tandem repeats (VNTR) where most loci have multiple alleles which can lead to a number of unresolvable bands (Jeffreys *et al.*, 1988).

DNA FINGERPRINTING PROTOCOL

Isolation of genomic DNA of *P. infestans*

1. Grow *P. infestans* in 9 cm Petri dishes in liquid rye medium (Caten and Jinks, 1968) inoculated with spores in the dark at 18°C until the plates are covered with mycelium. This will last about 10 days.

2. Harvest the mycelium, using a Büchner funnel, dry it with paper tissues and lyophilize overnight. Mycelium can be stored at −20°C. From one Petri dish approximately 150 mg of lyophilized mycelium can be obtained. Grind the lyophilized mycelium in liquid nitrogen in a mortar to a fine powder. Neither the mycelium nor the fine powder should be thawed.

3. Transfer the fine powder to a tube and add approximately 7 ml g^{-1} mycelium of a prewarmed (55°C) mixture (0.6:1) of H$_2$O-saturated phenol and extraction buffer (1:2:2 mixture of 5 × RNB : 12% PAS (w/v): 2% TNS (w/v). 5 × RNB is 200 mM Tris-HCl ph 8.5; 250 mM NaCl; 50 mM EGTA (Sigma, No. E4378). PAS is 4-aminosalicylic acid (Sigma, No. A3505) freshly prepared. TNS is triisopropyl-naphthalene sulphonic acid (Brunschwig 37035) freshly prepared). Incubate the mixture in a 55°C waterbath for 2 min and shake it every 10 s.

4. Add ¼ volume of chloroform/isoamyl alcohol (24:1) and incubate the mixture again at 55°C for 2 min and shake it every 10 s.

5. Centrifuge for 10 min (4000g) and transfer the aqueous phase to a new tube.

6. Add 1 volume of phenol/chloroform/isoamyl alcohol (25:24:1) to the aqueous phase and shake the mixture vigorously for 1 min.

7. Centrifuge for 10 min (4000g) and transfer the aqueous phase to a new tube.

8. Add one volume of chloroform/isoamyl alcohol (24:1) to the aqueous phase and shake the mixture vigorously for 1 min.

9. Centrifuge for 10 min (4000g) and transfer the aqueous phase to a new tube.

10. Repeat steps 8 and 9.

11. Add 0.6 volume of 2-propanol to the aqueous phase. Mix to precipitate the DNA. Collect the DNA in a pellet by centrifuging for 10 min (4000g).

12. Discard the supernatant, dry the inside of the tubes with a paper tissue and air-dry the pellets for 10 minutes.

13. Resuspend the pellets in the appropriate volume T$_{10}$E$_1$ (10 mM Tris-HCl (pH 8.0); 1 mM EDTA) containing RNase A (20 μg ml^{-1}). When mycelium from one Petri dish is used as starting material (± 150 mg), the pellet should be resuspended in 300 μl. Do

it carefully to avoid breakage of the DNA. Incubate at 37°C for 15–30 min. Pellets must be completely dissolved.

14. Estimate the DNA concentration by loading 5 μl of the DNA solution on a 0.7% agarose gel. Use a DNA solution of which the concentration is known (e.g. via spectrophotometric measurements) as a standard. Using this method, between 250 and 500 μg DNA per gram lyophilized mycelium can be obtained.

Restriction endonuclease digestion gel-electrophoresis, Southern blotting and hybridization

1. Digest 5 μg of genomic DNA with 20 U of the restriction enzyme *Eco*RI according to the manufacturer's instructions.
2. Size fractionate the restricted DNA on a 0.8% agarose gel for 14–16 h at 40 volts (600 V h). (Agarose gel electrophoresis according to Sambrook *et al.*, 1989.)
3. Transfer the DNA from the gel to a nylon hybridization membrane using 0.4 N NaOH as transfer buffer. We routinely use Hybond N$^+$ membranes from Amersham. Perform the capillary transfer according to the instructions of the manufacturer of the membrane and Sambrook *et al.* (1989).
4. Prehybridize the membrane for 2 h at 65°C in 10 ml hybridization solution (0.5 M NaH$_2$PO$_4$/Na$_2$HPO$_4$, pH 7.2; 1 mM EDTA; 7% SDS) in a rotating hybridization bottle.
5. Add a radioactive labelled DNA probe which can be prepared via the random primer labelling method described by Feinberg and Vogelstein (1983). Routinely we use 25 ng of insert DNA of the plasmid containing RG-57 and we label it with [a-^{32}P]dATP.
6. Hybridize overnight at 65°C.
7. Wash the membrane three times for 20 min each at 65°C in 2 × SSC, 0.5% SDS and three times 20 min in 0.5 × SSC, 0.1% SDS.
8. Expose the membrane at −80°C for 1–3 days to Kodak XOmat S film backed with an intensifier screen.
9. The membrane can be rehybridized with another probe after removal of the probe. This can be done by incubating the membrane in 0.4 N NaOH at 42°C for 30 min according to the manufacturer's instructions.

RESULTS

A detailed study on the spread and mode of reproduction of *P. infestans* isolates has been performed using RG-57 DNA fingerprinting to identify *P. infestans* isolates. In 1989 153 isolates were collected from 14 fields allocated over 6 regions in The Netherlands. Among these isolates 35 different RG-57 genotypes were identified. One genotype was particularly widespread. Half of the 153 isolates had this genotype and they were found in ten fields among five of the six regions sampled (Drenth *et al.*, 1993). Of the 35 genotypes 31 (89%) were found in only one field. DNA fingerprinting showed that the current Dutch *P. infestans* population represents a subpopulation of the population present in central Mexico. The hybridizing fragments in the DNA of these isolates collected in The Netherlands had all been detected before in the DNA of isolates from Mexico (Goodwin

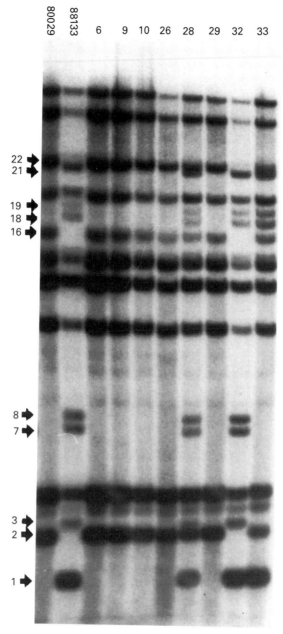

Fig 31.1.   RG–57 genotypes of *P. infestans* isolates. Autoradiograph of a Southern blot containing genomic DNA of *P. infestans* isolates and hybridized with probe RG–57. Numbers on the left indicate hybridizing fragments which are polymorphic in the two isolates that were used for a sexual crossing *in vitro*, 80029 and 88133. The numbers correspond to those used by Goodwin *et al.* (1992a,b). Isolate 80029 is homozygous for fragments 2 and 22 and heterozygous for fragment 16. Isolate 88133 is homozygous for fragments 1 and 21 and heterozygous for fragments 3, 7, 8, 18 and 19. Therefore, hybrid progeny should always possess fragments 1, 2, 21, 22 and a variable number of the heterozygous fragments from either one of the parents. Isolates 28 and 33 are hybrids resulting from sexual reproduction. Isolates 6, 9, 10, 26 and 29 and isolate 32 are asexual progeny from 80029 and 88133, respectively.

et al., 1992b; Drenth *et al.,* 1993). The amount of genetic diversity found in The Netherlands is consistent with the hypothesis that sexual reproduction generates new genotypes and occasional very fit isolates become widespread due to rapid asexual reproduction.

The high level of genetic diversity in the Dutch *P. infestans* population suggests that *P. infestans* reproduces sexually in the field. To find evidence for this suggestion the ability of *P. infestans* to form oospores in infected potato tissue was studied. In the past it has been shown that oospores can be formed in potato and tomato plants under greenhouse conditions (Frinking *et al.,* 1987) and in the field (unpublished), and viable progeny has been obtained from joining Dutch *P. infestans* isolates *in vitro* (Goodwin *et al.,* 1992a). However, the inoculum potential and the conditions for survival of oospores are unknown. In order to determine this, two isolates of opposite mating type and with distinguishable DNA fingerprint patterns, were inoculated on potato leaves to allow abundant oospore formation. These oospores appeared to be able to germinate and they gave rise to pathogenic progeny. The progeny were the result of sexual reproduction as was shown by DNA fingerprinting (Fig. 31.1). Leaves containing oospores were transferred to soil and subsequently potato tubers were planted in this soil. Infection of tubers and leaves of potato plants occurred and we are now in the process of determining whether oospores are responsible for the infection. Currently we are performing field experiments in which potato plants are inoculated using the same two isolates. These experiments will allow us to determine unambiguously the potential of *P. infestans* to survive and propagate by means of oospores in the soil under field conditions.

The advantages of DNA fingerprinting are obvious with regard to the number of loci which can be analysed in one assay. Probe RG-57 provides a sensitive tool for unambiguous identification of individual genotypes which is essential to perform population genetics on organisms with a potential mixed asexual/sexual life cycle. In sexual crosses selfed progeny can be distinguished from hybrid progeny. Additional uses for DNA fingerprinting will be in sorting out heterokaryosis, parasexuality, ploidy variation and the construction of molecular genetic linkage maps.

ACKNOWLEDGEMENT

We thank Dr Stephen B. Goodwin from the Department of Plant Pathology, Cornell University, Ithaca NY, USA, for providing probe RG-57.

REFERENCES

Carter, D. A., Archer, S. A., Buck, K. W., Shaw, D. S. and Shattock, R. C. (1990) Restriction fragment length polymorphisms of mitochondrial DNA of *Phytophthora infestans. Mycological Research* 94, 1123–1128.

Caten, C. E. and Jinks, J. L. (1968) Spontaneous variability of single isolates of *Phytophthora infestans.* I. Cultural variation. *Canadian Journal of Botany* 46, 329–348.

Drenth, A., Goodwin, S. B., Fry, W. E. and Davidse, L. C. (1993) Genotypic diversity of *Phytophthora infestans* in The Netherlands revealed by DNA polymorphisms. *Phytopathology* (in press).

Feinberg, A. P. and Vogelstein B. (1983) A technique for radiolabeling DNA restriction endonuclease fragments to high specific activity. *Analytical Biochemistry* 132, 6–13.

Frinking, H. D., Davidse, L. C. and Limburg, H. (1987) Oospore formation by *Phytophthora infestans* in host tissue after inoculation with isolates of opposite mating type found in the Netherlands. *Netherlands Journal of Plant Pathology* 93, 147–149.

Goodwin, S. B. (1991) DNA polymorphisms in *Phytophthora infestans*: the Cornell Experience. In: Lucas, J. A., Shattock, R. C., Shaw, D. S., Cooke, R. (eds) *Phytophthora*. Cambridge University Press, Cambridge, pp. 256–271.

Goodwin, S. B., Drenth, A. and Fry, W. E. (1992a) Cloning and genetic analyses of two highly polymorphic, moderately repetitive nuclear DNA probes from *Phytophthora infestans*. *Current Genetics* 22, 107–115.

Goodwin, S. B., Spielman, L. J., Matuszak, J. M., Bergeron, S. N. and Fry, W. E. (1992b) Clonal diversity and genetic differentiation among *Phytophthora infestans* populations in northern Mexico. *Phytopathology* 82, 955–961.

Hohl, H. R. and Iselin, K. (1984) Strains of *Phytophthora infestans* with A2 mating type behavior. *Transactions of the British Mycological Society* 83, 529–530.

Hooker, W. J. (ed.) (1981) Compendium of potato diseases. *American Phytopathological Society*. St Paul. MN.

Jeffreys, A. J., Royle, N. J., Wilson, V. and Wong, Z. (1988) Spontaneous mutation rates to new length alleles at tandem-repetitive hypervariable loci in human DNA. *Nature* 332, 278–281.

Maynard Smith, J. (1971) What use is sex? *Journal of Theoretical Biology* 30, 319–335.

Newhouse, J. R., Tooley, P. W., Smith, O. P. and Fishel, R. A. (1992) Characterization of double-stranded RNA in isolates of *Phytophthora infestans* from Mexico, the Netherlands, and Peru. *Phytopathology* 82, 164–169.

Rullich, G. and Schöber, B. (1988) Auftreten und Verbreitung der A2 Paarungstyps von *Phytophthora infestans* (Mont.) de Bary in der Bundesrepublik Deutschland. *Kartoffelbau* 39, 244–246.

Sambrook, J., Fritsch, E. F. and Maniatis, T. (eds) (1989) *Molecular Cloning: A Laboratory Manual*, 2nd edn. Cold Spring Harbor Laboratory Press, Cold Spring Harbor, New York.

Schöber, B. (1983) Definition und Auftreten der Pathotypen von *Phytophthora infestans* (Mont.) de Bary. *Kartoffelbau* 34, 156–158.

Shaw, D. S. (1983) The cytogenetics and genetics of *Phytophthora*. In: Erwin, D. C., Bartnicki-Garcia, S. and Tsao, P. H. (eds) *Phytophthora, its Biology, Taxonomy, Ecology and Pathology*. APS Press, pp. 81–94.

Spielman, L. J., McMaster, B. J. and Fry, W. E. (1989) Dominance and recessiveness at loci for virulence against potato and tomato in *Phytophthora infestans*. *Theoretical and Applied Genetics* 77, 832–838.

Spielman, L. J., Sweigard, J. A., Shattock, R. C. and Fry, W. E. (1990) The genetics of *Phytophthora infestans*: segregation of allozyme markers in F2 and backcross progeny and the inheritance of virulence against potato resistance R2 and R4 in F1 progeny. *Experimental Mycology* 14, 57–69.

Spielman, L. J., Drenth, A., Davidse, L. C. and Sujkowski, L. J., Gu, W. K., Tooley, P. W. and Fry, W. E. (1991) A second world–wide migration (and population displacement) of the potato late blight fungus? *Plant Pathology* 40, 422–430.

Tooley, P. W., Fry, W. E. and Villarreal Gonzalez, M. J. (1985) Isozyme characterization of sexual and asexual *Phytophthora infestans* populations. *Journal of Heredity* 76, 431–435.

Weir, B. S. (1992) Population genetics in the forensic DNA debate. *Proceedings of the National Academy of Sciences of the United States of America* 89, 11654–11659.

Williams, J. G. K., Kubelik, A. R., Livak, K. J., Rafalski, J. A. and Tingey, S. V. (1990) DNA polymorphisms amplified by arbitrary primers are useful as genetic markers. *Nucleic Acids Research* 18, 6531–6535.

Zischler, H., Schäfer, R. and Epplen, J. T. (1989) Non-radioactive oligonucleotide fingerprinting in the gel. *Nucleic Acids Research* 17, 4411.

32 Specificity of Monoclonal Antibodies Raised to *Pythium aphanidermatum* and *Erysiphe pisi*

A. J. Mitchell, A. J. Mackie, A. M. Roberts,
K. A. Hutchison, M. T. Estrada-Garcia, J. A. Callow
and J. R. Green

School of Biological Sciences, University of Birmingham, PO Box 363, Birmingham B15 2TT, UK.

Background

Many plant pathogens are classified by morphological and cultural characteristics (Van Der Plaats-Niterink, 1981). However, the methods used can have limitations for taxonomic identification, especially for those species showing similar morphology. Other techniques that have been employed to study taxonomic relationships include serological analysis (Krywienczyk and Dorworth, 1980) and the study of protein and DNA profiles (Clare *et al.*, 1968; Crowhurst *et al.*, 1992). Such techniques may prove time consuming however, thus necessitating simpler and more rapid methods for species identification. Recently, monoclonal antibodies (MAbs) have successfully been used to detect isolate-, species- and genus-specific components on the surface of certain fungi (Hardham and Suzaki, 1986; Dewey, 1992; Pain *et al.*, 1992). Such antibodies can be used for the identification of pathogenic fungi and for studying taxonomic relationships.

We are currently using MAbs to study fungal cell surface components and specialized infection structures formed in plant–pathogen interactions caused by the fungi *Pythium aphanidermatum* and *Erysiphe pisi*. This chapter describes the properties of three MAbs (PA1, PA2 and PA3) raised against *P. aphanidermatum* (Estrada-Garcia *et al.*, 1989) and four MAbs (UB7, UB8, UB10 and UB11) selected from the studies on *E. pisi* (Mackie *et al.*, 1991; Callow *et al.*, 1992; Green *et al.*, 1992). Some of these MAbs bind to specific cell types and in addition they demonstrate interesting specificities when tested on a wide range of higher plant and fungal species, although the MAbs were not selected for these properties when initially screened. These MAbs are therefore potentially useful for identification purposes and taxonomic studies.

Production and screening of monoclonal antibodies

Antibodies to *P. aphanidermatum* were raised by immunizing BALB/c mice (6–10 weeks old) with zoospores that had been fixed by the addition of an equal volume of 4% (w/v) glutaraldehyde in 100 mM Pipes buffer, pH 7.1, for 15 min (Hardham *et al.*, 1986; Estrada-Garcia *et al.*, 1989). Fixation was necessary to prevent encystment although inevitably a small percentage of cysts (approx. 2–3%) were present. Following fixation, zoospores were washed three times by centrifugation (650 g for 10 min) and resuspension in PBS. The number of zoospores was counted and, if necessary, the concentration was adjusted prior to intraperitoneal injection with 300 ml suspension of zoospores at $1.5–3.0 \times 10^7$ cells cm^{-3} in PBS into each mouse. A second intraperitoneal injection 4 weeks later was followed by a final boost immunization (1.5×10^7 zoospores in 200 ml PBS) which was given intravenously 4 days before fusion.

Antibodies to *E. pisi* were raised by immunizing BALB/c mice (6–8 weeks old) with isolated haustorial complexes (HCs) as described by Mackie *et al.* (1991). Two mice were immunized intraperitoneally with approx. 10^6 HCs suspended in 200 ml of PBS. A second intraperitoneal injection 2 weeks later preceded a final intravenous injection given 4 months after the boost immunization, 4 days before fusion.

Fusions between immune spleen cells from both sets of immunizations described above and NSO myeloma cells, and the subsequent cloning of hybridomas were performed as described by Galfre and Milstein (1985).

Tissue culture supernatants (TCS) from growing hybridoma cell lines produced from immunizations with zoospores and cysts of *P. aphanidermatum* were screened by indirect immunofluorescence (IIF). This method was also used to screen TCSs produced from immunizations with isolated HCs along with enzyme-linked immunosorbent assay (ELISA).

PROTOCOL

Indirect immunofluorescence (IIF)

Materials

1. Multiwell microscope slides (Flow Laboratories, Rickmansworth, UK).
2. Fixative solutions:
   A – 0.4% (w/v) glutaraldehyde, 8% (w/v) formaldehyde in 100 mM Pipes, pH 7.1.
   B – 8% (w/v) *p*-formaldehyde in 0.03 M Pipes, pH 7.5 containing 0.01 M CaCl$_2$
3. Wash buffer – phosphate buffered saline (PBS).
4. Rabbit antimouse immunoglobulins conjugated to fluorescein isothiocyanate (RAM-FITC, Dakopatts, UK.) diluted 1 in 20 with blocking solution: PBS containing either 10% (v/v) goat serum or 0.5 mg ml^{-1} bovine serum albumin (BSA).
5. Mounting medium: glycerol (90% v/v) containing 100 mg *p*-phenylenediamine dissolved in 10 ml PBS.

Method

1. For screening purposes IIF is performed on fixed cells. To a suspension of zoospores and cysts of *P. aphanidermatum* (prepared as described in Estrada-Garcia *et al.*, 1989) add an equal volume of fixative solution A and incubate for 15 min. For isolated HCs (prepared as described in Mackie *et al.*, 1991; use approx. 2×10^6 ml^{-1}) add an equal volume of fixative solution B and incubate at 4°C for 1 h.

2. After fixation wash the cell suspension by centrifugation (11,600 g for 15 s) and gentle resuspension of the pellet in PBS. Repeat twice before adjusting the final density of the cell suspension to 5×10^5 cells ml^{-1}.

3. Coat each of the wells of the multiwell slides with 10 μl aliquots of fixed cells and allow the slides to dry at room temperature for 1 h.

4. Rinse each slide twice by immersion in a Petri dish containing PBS and allow to dry slightly before pipetting 10 μl aliquots of TCS on to each well. Place slides in a humid chamber and incubate with the primary antibody for 45 min.

5. Rinse each slide three times with PBS before adding 10 μl aliquots of RAM-FITC in blocking solution to each well. Incubate slides, again placed in a humid chamber, for 45 min.

6. Rinse the slides twice with PBS followed by two washes with distilled water. Finally, after allowing the slides to dry slightly, pipette a drop of mounting medium on to each well and place a coverslip on to each slide. Sample fluorescence is then examined under a UV microscope.

ELISA

Materials

1. Antigens derived from fungal mycelia and plant leaves are prepared by homogenization of samples in a mortar and pestle precooled with liquid nitrogen, followed by addition of PBS containing 1 mM PMSF and storage at −70°C. After thawing, samples are homogenized using a Dounce hand homogenizer before use (Pain *et al.*, 1992).

2. Nunc maxisorp microtitre plates (GIBCO BRL, Life Technology, Scotland).

3. Poly-L-lysine (10 mg ml^{-1} in PBS).

4. Washing solution – PBS containing 0.05% (v/v) Tween 20.

5. Fixative solution – 0.5% (w/v) glutaraldehyde in PBS.

6. Blocking solution – 10 mg ml^{-1} BSA, 100 mM glycine and 0.02% (w/v) sodium azide in PBS.

7. Alkaline phosphatase-conjugated rabbit antimouse immunoglobulins (AP-RAMIG; Dakopatts, UK) diluted 1 in 600 with 0.5 mg ml^{-1} BSA in PBS.

8. Substrate solution – 1 mg ml^{-1} *p*-nitrophenylphosphate freshly dissolved in 9.7% (v/v) diethanolamine buffer pH 9.8 (prepare and store in darkness before use).

9. Stop solution – 3 M NaOH.

Method

Incubate the microtitre plate with each of the following:

1. 100 μl per well of poly-L-lysine for 2 h.

2. 50 μl (2.5 mg) per well of antigen (homogenate of fungal mycelium or pea leaves). Incubate for 30 min on an orbital plate shaker (600 rpm) followed by the addition of 50 μl of fixative solution for 3 min.

Table 32.1. Binding of MAbs to zoospores and cysts of species other than *Pythium aphanidermatum.*

| Species | MAbs | | |
	PA1	PA2	PA3
Peronosporales			
Pythium aphanidermatum	+	+	+
Pythium acanthicum	–	–	–
Pythium catenulatum	–	–	–
Pythium coloratum	–	–	–
Pythium dissoticum	–	–	–
Pythium mamillatum	–	–	–
Pythium middletonii	–	–	–
Pythium salpingophorum	–	–	–
Phytophthora infestans	–	–	–
Phytophthora nicotianae	–	–	–
Saprolegniales			
Aphanomyces cochliodes	–	–	–
Saprolegnia declina	–	–	–
Saprolegnia ferax	–	–	–
Saprolegnia parasitica	–	–	–

Source: Estrada-Garcia *et al.,* 1989. MAbs binding, determined by indirect immunofluorescence (IIF), is recorded as follows:–, no binding; +, positive binding.

3. 200 μl of blocking solution overnight at 4°C.
4. 50 μl of TCS for 30 min on an orbital shaker (600 rpm).
5. 100 μl of AP-RAMIG for 30 min on an orbital shaker (600 rpm).
6. 100 μl of substrate solution. Incubate in the dark for 45 min.
7. 50 μl of stop solution.
8. When the reaction has been stopped read the absorbance values of the samples at 405 nm on an ELISA plate reader.

Between each of the steps wash the plate four times with 200 μl per well of PBS containing 0.05% (v/v) Tween 20. Flick the plate to remove the solution between

Table 32.2. Characteristics of the panel of MAbs probes raised to zoospores and cysts of *Pythium aphanidermatum.*

| MAbs | Immuno-globulin subclass | Zoospore | | | Adhesive cyst coat | Mature cyst wall | Molecular nature of antigen | Probable identity of epitope |
		AF	Body	PF				
PA1	IgM	++	++	++	–	–	75 kDa protein	protein
PA2	IgG1	++	–	–	–	–	?	?
PA3	IgG1	–	±	–	++	–	200 kDa glycoprotein	CHO

Source: Estrada-Garcia *et al.,* 1989, 1990a,b. MAbs binding, determined by indirect immunofluorescence (IIF), is recorded as follows:–, no binding; +, weak binding; ++, strong binding; ±, binding to patches of some cells. AF, anterior flagellum; PF, posterior flagellum.

washes. All incubations are performed at room temperature unless stated otherwise. A MAb (UBIM 22) raised against rat bone cells and known not to cross-react with fungal antigens was used as a negative control for IIF and ELISA assays.

RESULTS

From six fusions with mice immunized with zoospores and cysts from *P. aphanidermatum*, a total of eight antibodies showed interesting specificities in initial screening tests by IIF (Estrada-Garcia *et al.*, 1989). Three of these MAbs (designated PA1, PA2 and PA3) were specific for *P. aphanidermatum* (Table 32.1). These MAbs also demonstrated cell-specific binding to components of the zoospore cell surface (i.e. PA1 and PA2) or the adhesive cyst coat (i.e. PA3; Table 32.2).

MAbs raised to pea powdery mildew HCs were screened on HCs by IIF and on fungal and pea membranes by ELISA (Mackie *et al.*, 1991). Three MAbs (i.e. UB8, UB10 and UB11 in Table 32.3) recognized glycoproteins specifically located within the HC, thus demonstrating molecular differentiation of the membranes within this structure. These antibodies bind specifically to the haustorium formed by *E. pisi* and show no cross-reaction with related infection structures formed by several other fungi tested (Table 32.3).

Table 32.3. Characteristics of the panel of MAb probes raised to isolated haustorial complexes of *Erysiphe pisi.*

MAbs	Molecular nature of antigen (epitope)	Antigen location	Binding to infection structures formed in various fungal-(plant) interactions				
			E. pisi (pea)	*E. graminis* (wheat)	*E. cichoracearum* (antirrhinum)	*Peronospora viciae* (pea)	*Colletotrichum lindemuthianum* (bean)
UB7 IgG2a	62 kDa N-linked gp (CHO)	Surface hyphae and haustoria (pm/ wall)	+	+	+	−	+
UB8 IgG1	62 kDa N-linked gp (protein)	Haustorium-specific (hpm)	+	−	−	−	−
UB10 IgG1	45 kDa gp (CHO)	Haustorium-specific (hpm)	+	−	−	−	−
UB11 IgG1	250 kDa glyco-protein (CHO)	Haustorium-specific (ehm)	+	−	−	−	−

Source: Mackie *et al.*, 1991; Callow *et al.*, 1992. pm: plasma membrane; hpm: haustorial plasma membrane; ehm: extrahaustorial membrane; CHO: carbohydrate; gp: glycoprotein.

Another MAb (designated UB7) recognized a major cell wall and plasma membrane glycoprotein of haustoria and the surface mycelium (Table 32.3). This antibody shows cross-reaction with other infection structures studied (Table 32.3).

MAb UB7 was tested for cross-reaction with a wide range of fungal species and higher plants (leaves) using ELISA (Table 32.4). Results showed that UB7 bound to some, but not all species of fungi containing chitin in their cell walls and showed no

Table 32.4. Binding of MAbs UB7 to a wide range of fungi.

Species	MAbs UB7
Ascomycetes	
Erysiphe pisi	++
Botrytis allii	+++
Botrytis cinerea	+++
Penicillium corymbiferum	+++
Penicillium oxalicum	+++
Verticillium lecanii	−
Fusarium oxysporum f.sp. *lycopersici*	−
Fusarium culmorum	−
Fusarium graminearum	−
Saccharomyces cerevisiae	−
Aspergillus glaucus	+++
Aspergillus quadrilineatus	++
Ascobolus crenulatus	++
Trichoderma viride	+
Neurospora crassa	+
Zygomycetes	
Mucor hiemalis	−
Rhizopus stolonifera	+++
Phycomyces blakesleeanus	−
Rhizopus sexualis	++
Basidiomycetes	
Tricholoma nudum	−
Rhizoctonia solani	+++
Sporobolomyces roseus	−
Phellinus ferreus	−
Oomycetes	
Pythium ultimum	−
Pythium oligandrum	−
Saprolegnia parasitica	−
Phytophthora cactorum	−
Phytophthora infestans	−
Phytophthora erythroseptica	−
Deuteromycetes	
Cladosporium herbarum	+++
Cladosporium cucumerinum	+++
Phoma betae	+++
Phoma exigua	++

MAbs binding, determined by ELISA, is recorded as follows: −, OD <0.19; +, OD 0.2–0.49; ++, OD 0.5–0.99; +++, OD > 1.0.

cross-reaction with fungal species lacking this polysaccharide (e.g. Oomycetes). UB7 did not bind to any of the higher plants tested (pea, bean, rice, wheat, tobacco, oat; data not shown). Western blotting experiments with mycelial homogenates showed UB7 bound to different sets of proteins within different species (data not shown).

INTERPRETATION

Overall, results from these studies suggest that many of the MAbs selected for studies on plant–pathogen interactions also have interesting specificities when tested on a range of fungi. Antibodies which have been selected because they show binding to specific cell types (PA1, PA2, PA3 for *P. aphanidermatum*; UB8, UB10, UB11 for *E. pisi*) have turned out to be species-specific and therefore have potential for identification purposes. MAb UB7 which shows more general binding does not cross-react with higher plant antigens, but recognizes many fungal species (except for the Oomycetes). UB7 may be useful for the detection and/or confirmation of a fungal pathogen, suspected to be present in plant material.

The advantage of using the IIF method in the preliminary stages of antibody screening has been highlighted in these studies. Screening TCSs from hybridoma cell lines by IIF and selecting those which show binding to specific cell types may increase the possibility of obtaining antibodies which demonstrate species-specific binding. Immunization with specific cell types rather than crude homogenates may also increase the percentage of species-specific antibodies obtained. Clearly the limitation of MAbs which show restricted binding to specific cell types is that it may not be possible to ensure that the relevant structures are present in a sample which is being used in diagnostic tests.

If we consider the methods used in these studies with respect to their application for diagnostic use by untrained personnel in the field, it is obvious that the IIF assay would not be suitable for commercial development. However, MAb-based ELISAs or dipstick assays may, with further testing and development, show potential for commercial use. With these latter methods the level of sensitivity at which reliable and reproducible results can be obtained requires further investigation. For instance, soil infected with *P. aphanidermatum* may only comprise one or two zoospores per cubic centimetre of soil, and this is likely to be below the level of detection for an ELISA or dipstick assay. Thus, the level of sensitivity provided by these assays may ultimately dictate their potential for commercial development.

REFERENCES

Callow, J. A., Mackie, A. J., Roberts, A. M. and Green, J. R. (1992) Evidence for molecular differentiation in powdery mildew haustoria through the use of monoclonal antibodies. *Symbiosis* 14, 237–246.

Clare, B. G., Flentje, N. T. and Arkinson, M. R. (1968) Electrophoretic patterns of oxidoreductases and other proteins as criteria in fungal taxonomy. *Australian Journal of Biological Sciences* 21, 275–295.

Crowhurst, R. N., Hawthorne, B. T., Rikkerink, E. H. A. and Templeton, M. D. (1992) Differentiation of *Fusarium solani* f. sp. *curcurbitae* races 1 and 2 by random amplification of polymorphic DNA. *Current Genetics* 20, 391–396.

Dewey, F. M. (1992) Detection of plant invading fungi by monoclonal antibodies. In: Torrance, J. M. and Torrance, L. (eds) *Techniques for the Rapid Detection of Plant Pathogens*. Blackwell Scientific, Oxford, pp. 47–62.

Estrada-Garcia, M. T., Green, J. R., Booth, J. M., White, J. G. and Callow, J. A. (1989) Monoconal antibodies to cell surface components of zoospores and cysts of the fungus *Pythium aphanidermatum* reveal species-specific antigens. *Experimental Mycology* 13, 348–355.

Estrada-Garcia, M. T., Callow, J. A., and Green, J. R. (1990a) Monoclonal antibodies to the adhesive cell coat secreted by *Pythium aphanidermatum* zoospores recognise $200 \times 10^3 \, M_r$ glycoproteins stored within large peripheral vesicles. *Journal of Cell Science* 95, 199–206.

Estrada-Garcia, M. T., Ray, T. C., Green, J. R., Callow, J. A. and Kennedy, J. F. (1990b) Encystment of *Pythium aphanidermatum* zoospores is induced by root mucilage polysaccharides, pectin and a monoclonal antibody to a surface antigen. *Journal of Experimental Botany* 41, 693–699.

Galfre, G. and Milstein, C. (1985) Preparation of monoclonal antibodies: Strategies and procedures. In: Langone, J. J. and Van Vunakis, H. (eds) *Methods in Enzymology*, vol. 73. Academic Press, Orlando, pp. 1–46.

Green, J. R., Mackie, A. J., Roberts, A. M. and Callow, J. A. (1992) Molecular differentiation and development of the host–parasite interface in powdery mildew of pea. In: Callow, J. A. and Green, J. R. (eds) *Perspectives in Plant Cell Recognition*. Society for Experimental Biology Seminar Series 48. Cambridge University Press, pp. 193–212.

Hardham, A. R. and Suzaki, E. (1986) Encystment of zoospores of the fungus *Phytophthora cinnamomi* is induced by a specific lectin and a monoclonal antibody binding to the cell surface *Protoplasma* 133, 165–173.

Krywienczyk, J. and Dorworth, C. E. (1980) Serological relationships of some species in the genus *Pythium*. *Canadian Journal of Botany* 58, 1412–1417.

Mackie, A. J., Roberts, A. M., Callow, J. A. and Green, J. R. (1991) Molecular differentiation in pea powdery-mildew haustoria. Identification of a 62-kDa N-linked glycoprotein unique to the haustorial plasma membrane. *Planta* 183, 399–408.

Pain, N. A., O'Connell, R. J., Bailey, J. A. and Green, J. R. (1992) Monoclonal antibodies which show restricted binding to four *Colletotrichum* species: *C. lindemuthianum*, *C. malvarum*, *C. orbiculare* and *C. trifolii*. *Physiological and Molecular Plant Pathology* 40, 111–126.

Van Der Plaats-Nitterink, A. J. (1981) *Monograph of the Genus* Pythium. *Studies in Mycology* No. 21. Centralbureau voor Schimmelcultures, Baam.

33 Allele-specific Oligonucleotides and their Use in Characterization of Resistance to Benomyl in *Venturia inaequalis*

H. KOENRAADT[1] AND A. L. JONES

Department of Botany and Plant Pathology, Michigan State University, East Lansing, Michigan 48824, USA.
[1] *Present address: NAKG, PO Box 27, 2370 AA Roelofarendsveen, The Netherlands.*

BACKGROUND

Venturia inaequalis (Cooke) Wint., the causal organism of apple scab, is an economically important pathogen of apples in several areas in the world. Benomyl, a benzimidazole compound, was introduced in the early 1970s and gave excellent control of apple scab. However, in the third year of extensive and exclusive applications in apple orchards in Michigan, control of apple scab failed due to the development of resistance to benomyl (Jones and Walker, 1976). Benomyl-resistant strains of *V. inaequalis* have been isolated in many apple growing regions worldwide. In addition to sensitive (S) strains, four classes of resistant strains (low, LR; medium, MR; high, HR; and very highly resistant, VHR) were established based on their growth response on media amended with benomyl (Katan *et al.*, 1983; Shabi *et al.*, 1983). Classical genetic analysis indicated that benomyl resistance was determined by a single gene in *V. inaequalis* (Katan *et al.*, 1983; Shabi *et al.*, 1983; Stanis and Jones, 1984).

In binding studies, β-tubulin from a benomyl-resistant strain of *Aspergillus nidulans* exhibited reduced affinity for carbendazim, the toxic conversion product of benomyl (Davidse and Flach, 1977). Nucleotide sequence analysis of the β-tubulin gene of benomyl-resistant strains of *Neurospora crassa*, and *A. nidulans* provided evidence that single base-pair mutations in this gene were responsible for resistance to benomyl (Orbach *et al.*, 1986; Jung and Oakley, 1990; Jung *et al.*, 1992). Characterization of the β-tubulin gene of sensitive and benomyl-resistant strains showed that single base-pair mutations were also present in benomyl-resistant field strains of *V. inaequalis*. Codon 198, which encodes glutamic acid in a sensitive strain, was converted to a codon for alanine in a VHR strain, to a codon for lysine in a HR strain, or to a codon for glycine in a MR strain. Codon 200 for phenylalanine was converted to a codon for tyrosine in a second MR strain (Koenraadt *et al.*, 1992). Consecutive transformation experiments, in which *N. crassa* was used as a model fungus, showed that mutations in codon 198 and 200 were directly responsible for resistance to benomyl (Koenraadt and Jones, 1993). To characterize point mutations in additional strains of *V. inaequalis* we utilized allele-specific oligonucleotide (ASO) analysis. ASO analysis allows the rapid detection of point mutations that confer resistance to benomyl in *V. inaequalis* and provides an invaluable alternative to conventional and laborious sequence analysis by the dideoxy chain termination method (Koenraadt *et al.*, 1992).

Allele-specific oligonucleotide (ASO) analysis

In this method, PCR is used to amplify the appropriate DNA sequence and then ASO analysis is used to detect single base-pair mutations. Under conditions of high stringency, ASO probes with a single mismatch, but not the probe with a complete match, will be removed from their complementary sequence due to a reduced stability of the hybridization complex.

PROTOCOL

ASO analysis overview

Materials

1. Test material: mycelium from single-spore isolates of *V. inaequalis* or leaves and fruits with scab lesions.
2. DNA extraction: lysis buffer (100 mM LiCl, 50 mM Na$_2$EDTA, 1.0% SDS, 10 mM Tris-HCl, pH 7.4) and phenol.
3. PCR: two 22-mer oligonucleotides, 5'-CAAACCATCTCTGGCGAACACACG and 5'-TGGAGGACATCTTAAGACCACG.
4. Agarose electrophoresis buffer 1 × TBE (0.1 M Tris-HCl, 0.1 M boric acid, 0.02 mM EDTA, pH 8.3).
5. Four ASO probes (ASO[S-LR], ASO[MR], ASO[HR] and ASO[VHR]).
6. End-labelling ASO probes: [γ-^{32}P]ATP (Du Pont, Boston, MA) and T4 polynucleotide kinase (Promega Corporation, Madison, WI).
7. Nylon membrane (GeneScreen-Plus, Du Pont, Boston, MA) and a dot-blot manifold.
8. Prehybridization solution: 1 M NaCl, 50 mM Tris-HCl, pH 7.5, 10% dextran sulphate, 1% SDS, 0.2% Ficoll (MW 400,000), 0.2% polyvinylpyrrolidone (MW 400,000), 0.2% bovine serum albumin, 0.1% sodium pyrophosphate, and 0.25 mg ml^{-1} denatured salmon sperm DNA.
9. Washing buffer 2 × SSC (1 × SSC is 0.15 M NaCl, 0.015 M sodium citrate, pH 6.8).

Method

1. Extract total DNA from mycelium or from scab lesion.
2. Amplify β-tubulin DNA with PCR and analyse amplification product by agarose gel electrophoresis.
3. End-label ASO probes with [γ-^{32}P]ATP using T4 polynucleotide kinase to a minimum specific activity of 2 × 10^9 dpm pmol^{-1}.

4. Denature PCR amplified β-tubulin DNA (25 ng per sample) in 0.25 N NaOH for 10 min.

5. Apply denatured DNA to a membrane in dot-blot manifold.

6. Incubate dot blots in prehybridization solution.

7. Add end-labelled ASO probe to the prehybridization solution and incubate at 37°C for at least 4 h.

8. Wash dot blots three times for 15 min each in 2 × SSC buffer at room temperature.

9. Wash dot blots at high stringency three times for 2 min each in 2 × SSC buffer.

10. Expose dot blots to X-ray film for 0.5–2 days at −70°C.

ELUCIDATION OF THE PROTOCOL

Fungal strains

The single-spore isolates of *V. inaequalis* used in this study were from a large collection of field strains previously characterized in studies on the inheritance of resistance to benomyl and of negatively correlated cross-resistance to diethofencarb (Jones *et al.*, 1987; Shabi *et al.*, 1987). Each strain was designated as sensitive (S), low resistance (LR), medium resistance (MR), high resistance (HR), or very high resistance (VHR) to benomyl based on its growth response on media amended with benomyl or diethofencarb.

DNA isolation and polymerase chain reaction

Mycelium (10–100 mg wet weight) from cultures grown in potato dextrose broth (PDB) or on potato dextrose agar (PDA) (Difco Laboratories, Detroit, MI) was transferred to microcentrifuge tubes, frozen with liquid nitrogen, and then ground to a powder with disposable pellet pestles (Kontes, Morton Grove, IL). The mycelium was suspended by vortexing in 0.7 ml of lysis buffer (100 mM LiCl, 50 mM Na_2EDTA, 1.0% SDS, 10 mM Tris-HCl, pH 7.4) with 20 $\mu g\,ml^{-1}$ proteinase K (Boehringer Mannheim, Indianapolis, IN) and then incubated at 50°C for 1 h. The DNA was purified by phenol extraction and ethanol precipitation. The isolated DNA was then used as template in PCR (Erlich *et al.*, 1988). Two 22-mer oligonucleotides, 5′-CAAACCATCTCTGGCGAACACACG and 5′-TGGAGGACATCTTAAGACCACG, which were identical in sequence to codons 22 to 28 and complementary in sequence to codons 359 to 365 of the β-tubulin gene of *V. inaequalis*, respectively, were used as PCR primers to amplify a 1188-bp DNA sequence. PCR was performed and amplification products were analysed by agarose gel electrophoresis as described previously (Koenraadt *et al.*, 1992).

Allele-specific oligonucleotide probes

Four ASO probes for detecting allelic mutations in the β-tubulin gene of *V. inaequalis* were synthesized in the Macromolecular Structure Facility, Department of Biochemistry, Michigan State University, East Lansing. The 18-mer oligonucleotides were designated as ASO[S-LR], ASO[MR], ASO[HR], and ASO[VHR] probes based on the specificity of each probe (Fig. 33.1). Each ASO probe included the sequences of codons 198 and 200 because mutations in these codons were associated with resistance to benomyl in field strains of

			198		200			
ASO^{S-LR} 5′	C	TCT	GAC	GAG	ACA	TTC	TG	3′
ASO^{MR}			C	GAG	ACA	T<u>A</u>C	TGC	ATT	GA
ASO^{HR}	C	TCT	GAC	<u>A</u>AG	ACA	TTC	TG
ASO^{VHR}	C	TCT	GAC	G<u>C</u>G	ACA	TTC	TG

Fig 33.1. Sequences of allele-specific oligonucleotide (ASO) probes for *Venturia inaequalis*. The ASO probes for medium (MR), high (HR) and very high resistance (VHR) to benomyl differ from the *β*-tubulin DNA for sensitive (S) and low resistance (LR) strains by one nucleotide (underlined).

V. inaequalis (Koenraadt *et al.*, 1992). ASO probes were end-labelled with [γ-^{32}P]ATP (Du Pont, Boston, MA) using T4 polynucleotide kinase (Promega Corporation, Madison, WI) to a minimum specific activity of 2×10^9 dpm pmol^{-1}.

Allele-specific oligonucleotide analysis

PCR-amplified *β*-tubulin DNA (25 n per sample) was denatured in 0.25 N NaOH for 10 min and then applied to a nylon membrane (GeneScreen-Plus, Du Pont, Boston, MA) in a dot-blot manifold. The dot blots were incubated in prehybridization solution (1 M NaCl, 50 mM Tris-HCl, pH 7.5, 10% dextran sulphate, 1% SDS, 0.2% Ficoll (MW 400,000), 0.2% polyvinylpyrrolidone (MW 40,000), 0.2% bovine serum albumin, 0.1% sodium pyrophosphate, and 0.25 mg ml^{-1} denatured salmon sperm DNA) for 2 h at 65°C according to the manufacturer's procedure. An end-labelled ASO probe was then added to the prehybridization solution and incubated at 37°C for at least 4 h. The blots were washed three times for 15 min each in 2 × SSC buffer (1 × SSC is 0.15 M NaCl, 0.015 M sodium citrate, pH 6.8) at room temperature. A high stringency wash, three times for 2 min each in 2 × SSC buffer, was then used to remove ASO probes with a single base-pair mismatch from the blots. The optimum temperature for the high stringency wash with each probe was determined empirically. The dot blots were exposed to X-ray film for 0.5–2 days at −70°C.

RESULTS

DNA isolated from as little as 10 mg fresh weight of mycelium from broth or agar plate cultures was sufficient for the amplification of the 1188-bp target sequence of the *β*-tubulin gene using PCR. Each of the four ASO probes hybridized to the PCR amplified *β*-tubulin DNA of all strains of *V. inaequalis* after washing the dot blots at 50°C (Fig. 33.2A). After washing at 64°C, the ASO^{S-LR} probe hybridized only with the PCR amplified *β*-tubulin DNA of strains with S or LR phenotype, and not to DNA of strains with the MR, HR or VHR phenotype (Fig. 33.2B). The ASO^{MR}, ASO^{HR} and ASO^{VHR} probes exhibited allele-specific hybridization after washing at 61, 61 and 63°C, respectively (Fig. 33.2C–E). ASO analysis showed that codons for lysine and alanine at position

198 were always associated with the HR and VHR phenotypes, respectively; while base substitutions giving a tyrosine codon at position 200 were always associated with the MR phenotype. The independent selection of identical codon conversions in *β*-tubulin DNA of field strains from diverse geographic regions provides additional evidence that these mutations are the primary basis for resistance to benomyl. No mutation was detected in strains with the LR phenotype (Koenraadt and Jones, 1992). Until the molecular basis for the LR phenotype is established, it will be impossible to distinguish between S and LR strains using ASO analysis.

LIMITATIONS

ASO analysis involves the following steps: extracting DNA, amplifying the target sequence using PCR, and probing amplified DNA with allele-specific probes. A rapid and simple minipreparation extraction procedure was developed to reduce the chance of contaminating the extraction mixture with DNA that could serve as template during PCR amplification. Amplification of *β*-tubulin DNA by PCR obviated the need to obtain large amounts of mycelium for DNA extraction. After hybridizing the probe with amplified DNA at 37°C, dot blots were washed at an elevated temperature to remove probes with a single base-pair mismatch. The optimum wash temperature for removing non-specific probes varied for each ASO probe and was related to the G + C content of the probes.

With ASO analysis, the sequence must be known before PCR primers and ASO probes can be designed. Also, degeneracy in the genetic code interfered with the characterization of allelic mutations in the *β*-tubulin DNA of other plant pathogenic fungi. For example, none of the *V. inaequalis*-specific ASO probes hybridized with amplified *β*-tubulin DNA from *V. pirina* due to variations in the third position of three codons (Koenraadt and Jones, 1992). A set of species-specific probes could circumvent this limitation of ASO analysis.

ASO analysis is a rapid method for characterizing point mutations in *β*-tubulin alleles that confer resistance to benomyl in field strains of *V. inaequalis*. The method is sensitive and versatile. Both axenic mycelium and tissue from apple scab lesions can be assayed. For the direct characterization of alleles apple scab lesions were cut from leaves or fruits with a scalpel and individually transferred to microcentrifuge tubes. Isolation of DNA from the lesions, amplification of the target sequence by the PCR, and ASO analysis were done as described above. Because ASO analysis can be performed on material from individual lesions, as demonstrated in a previous study (Koenraadt and Jones, 1992), this method could be employed to identify and characterize benomyl resistance in pathogens that are difficult to culture and in obligate parasites such as powdery mildews. Provided that resistance is based on point mutations, ASO analysis might also assist in determining the potential variability of target sites for new pesticides with a site-specific mode of action. In conclusion, we found that ASO analysis was an easy, rapid method, and believe that it can be readily adapted to a range of tests to determine relevant single base-pair mutations in plant pathogenic organisms.

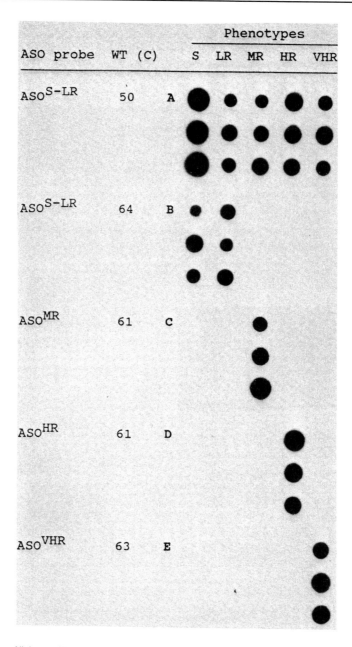

Fig 33.2. Allele-specific oligonucleotide analysis with amplified β-tubulin DNA of *Venturia inaequalis*. The ASO^S-LR probe hybridized to DNA on dot blots in panels A and B, and the blots were subsequently washed at a temperature (WT) of low stringency (A) or optimum stringency (B). The ASO^MR, ASO^HR and ASO^VHR probes were hybridized to DNA on the blots in panels C, D and E, respectively. The blots were subsequently washed at optimum stringency. Amplified β-tubulin DNA, from sensitive, low, medium, high and very high resistant strains was applied to each blot in column S, LR, MR, HR and VHR, respectively. S strains (from top to bottom): B7, MSU-18 and WC; LR strains: MINNS 118, MITCHELL 94–1 and MSU-20; MR strains: MAINE 4, MAINE 8 and SIS-16; HR strains: RI, I-23 and RH-4; and VHR strains: TU86R2, 75–26 and KV3C (from Koenraadt and Jones, 1992).

REFERENCES

Davidse, L. C. and Flach, W. (1977) Differential binding of methyl-2-yl carbamate to fungal tubulin as mechanism of resistance to this anti-mitotic agent in mutant strains of *Aspergillus nidulans*. *Journal of Cell Biology* 72, 173–193.

Erlich, H. A., Gelfand, D. H. and Saiki, R. S. (1988) Specific DNA amplification. *Nature* 331, 461–462.

Jones, A. L. and Walker, R. J. (1976) Tolerance of *Venturia inaequalis* to dodine and benzimidazole fungicides in Michigan. *Plant Disease Reporter* 60, 40–44.

Jones, A. L., Shabi, E. and Ehret, G. R. (1987) Genetics of negatively correlated cross-resistance to *N*-phenylcarbamate in benomyl-resistant *Venturia inaequalis*. *Canadian Journal of Plant Pathology* 9, 195–199.

Jung, M. K. and Oakley, B. R. (1990) Identification of an amino acid substitution in the *ben*A, beta-tubulin gene of *Aspergillus nidulans* that confers thiabendazole resistance and benomyl supersensitivity. *Cell Motility and Cytoskeleton* 17, 87–94.

Jung, M. K., Wilder, I. B. and Oakley, B. R. (1992) Amino acid alternations in the *ben*A (beta-tubulin) gene of *Aspergillus nidulans* that confer benomyl resistance. *Cell Motility and Cytoskeleton* 22, 170–174.

Katan, T., Shabi, E. and Gilpatrick, J. D. (1983) Genetics of resistance to benomyl in *Venturia inaequalis* from Israel and New York. *Phytopathology* 73, 600–603.

Koenraadt, H. and Jones, A. L. (1992) The use of allele-specific oligonucleotide probes to characterize resistance to benomyl in field strains of *Venturia inaequalis*. *Phytopathology* 82, 1354–1358.

Koenraadt, H. and Jones, A. L. (1993) Resistance to benomyl conferred by mutations in codon 198 or 200 of the beta-tubulin gene of *Neurospora crassa* and sensitivity to diethofencarb conferred by codon 198. *Phytopathology* 83, 850–854.

Koenraadt, H., Somerville, S. and Jones, A. L. (1992) Characterization of mutations in the beta-tubulin gene of benomyl-resistant field strains of *Venturia inaequalis* and other plant pathogenic fungi. *Phytopathology* 82, 1348–1354.

Orbach, M. J., Porro, E. B. and Yanofsky, C. (1986) Cloning and characterization of the gene for beta-tubulin from a benomyl-resistant mutant of *Neurospora crassa* and its use as a dominant selectable marker. *Molecular and Cellular Biology* 6, 2452–2461.

Shabi, E., Katan, T. and Marton, K. (1983) Inheritance of resistance to benomyl in isolates of *Venturia inaequalis* from Israel. *Plant Pathology* 32, 207–212.

Shabi, E., Koenraadt, H. and Dekker, J. (1987) Negatively correlated cross-resistance to phenyl-carbamate fungicides in benomyl-resistant *Venturia inaequalis* and *V. pirina*. *Netherlands Journal of Plant Pathology* 93, 33–41.

Stanis, V. F. and Jones, A. L. (1984) Genetics of benomyl resistance in *Venturia inaequalis* from North and South America, Europe and New Zealand. *Canadian Journal of Plant Pathology* 6, 283–290.

34 The Development of Immunological Techniques for the Detection and Evaluation of Fungal Disease Inoculum in Oilseed Rape Crops

D. Schmechel, H. A. McCartney and K. Halsey

Department of Plant Pathology, AFRC Institute of Arable Crops Research, Rothamsted Experimental Station, Harpenden, Herts. AL5 2JQ, UK.

Background

For crop protection systems to be environmentally acceptable while remaining economically efficient control measures should be taken only when the crop is at risk. In many crop disease epidemics, the pathogen can be well established before symptoms are visible and consequently control measures based on disease detection by direct observation may be less effective than they would have been had the disease been detected earlier. Early infection can be detected by monitoring the exposure of the crop to pathogen inoculum. For air-dispersed fungal pathogens this can only be done, at present, using spore traps, which are both labour- and time-intensive and require trained personnel to identify the pathogen spores. Novel methods based on the immunological identification of airborne spores could provide a more straightforward and efficient way of quantifying the risk of a crop to infection. Additionally, similar immunodetection techniques could be used to quantify the extent of infection.

Oilseed rape (*Brassica napus* L.) is subject to attack by several fungal pathogens which can cause damaging disease epidemics. Pathogens of particular importance are: *Phoma lingam, Botrytis cinerea, Sclerotinia sclerotiorum, Alternaria* spp. and *Pyrenopeziza brassicae*. These pathogens may be present at low levels throughout the growing season, but symptoms may not be noted until epidemics are well established. Thus, as noted above, easy to use quantitative methods for the early assessment, before symptoms develop, of potential epidemic development would be a valuable tool in rational disease control systems.

This chapter describes work on some of the problems encountered while developing immunological techniques for the detection of airborne inoculum of oilseed rape pathogens.

Principles and test development

Monoclonal antibodies

Monoclonal antibodies were chosen as detection agents for fungal spores and fungal mycelium because of their potential specificity. For this study monoclonal antibodies were raised against germinating spores of isolates of *Alternaria brassicae* and against mycelium of *Sclerotinia sclerotiorum*. For *Alternaria brassicae*, BALB/c mice were immunized with 0.4×10^6 spores suspended in $300 \mu l$ saline. Spores had been allowed to germinate overnight at room temperature. Mice were given five injections into the peritoneum. Hybridoma supernatants were screened by ELISA, using plates coated with spores which had been allowed to germinate overnight at room temperature. Five cell lines (IgM) were established, of which two (4F.3/1/1 and 4E3/3/2) showed low cross-reactivity against airborne inoculum of other *Alternaria* spp. and unrelated fungal species. For *Sclerotinia sclerotiorum*, BALB/c mice were given seven intraperitoneal injections with sonicated mycelium from scrapings of solid cultures suspended in saline (6 mg fresh weight in 0.5 ml, initial dose). Hybridoma culture supernatants were screened by ELISA. Plates were coated with the immunogen for 1 h at 37°C. Six cell lines (IgG$_1$) were established, of which two (17/3/a and 17/3/c) showed no cross-reactivity with *Botrytis cinerea* and *Pyrenopeziza brassicae*. A monoclonal antibody (BC-KH$_4$) that recognizes germinating spores of *Botrytis cinerea* was provided by Dr Molly Dewey, Department of Plant Science, University of Oxford.

Spore trapping

Several methods are used to assess concentrations of airborne spores (Bartlett and Bainbridge, 1978). Three types of trap are commonly used: suction traps, which collect spores by impacting them on sticky surfaces; liquid impingers, which collect spores in liquid; and roto-rod traps, which collect spores on rotating arms. For both suction traps and liquid impingers substantial sample processing, such as spore removal or sample concentration, would be required before antibodies could be used to detect spores. This may lead to loss of antigen, and consequent loss of sensitivity and accuracy of detection. Immunofluorescence techniques may be useful for identifying spores directly on suction trap samples, but their use would still require examination of the sample by microscope. Suction traps and liquid impingers need vacuum pumps and this often makes them bulky and expensive. In contrast roto-rod traps are relatively small, inexpensive and easy to use. A roto-rod trap was therefore adapted by attaching ELISA strip wells (Nunc-Immuno-Modules) to the rotating arms (Fig. 34.1). This means that the amount of sample processing before assessment can be minimized as spores are collected directly in the ELISA wells.

Tests, using *Lycopodium* spores, were done in a wind tunnel to find the best method of aligning the ELISA strips so that spores were impacted on the base of the wells (Fig. 34.1). The strips were attached to the supporting arms by rubber bands so that they could easily be removed for processing. During operation the arms rotated at about 2800 rpm, equivalent to an air sampling rate of about $350 \, l \, min^{-1}$.

Analysis of spore sample

After sampling, spores were incubated in the wells overnight to promote germination and subjected to an indirect ELISA the following day. All tests were done using the alkaline phosphatase system. The alkaline phosphatase system was chosen because removal of endogenous alkaline phosphatase activity was found to be easier than with endogenous peroxidase activity. Results were expressed as relative optical density, that is the sample optical density was divided by the background optical density (i.e. antigen, plain culture supernatant and secondary antibody). This was done to compare different samples taken in different experiments. A relative optical density of greater than 2.5 was considered positive.

It was found that the fungal spores were too large to be efficiently retained in ELISA wells during the washing steps. This resulted in antigen loss because the antibodies used recognize surface-bound antigens. Precoating ELISA wells with poly-L-lysine (Mazia *et al.*, 1975, Heusser *et al.*, 1981; Huang *et al.*, 1983) did not improve spore retention. However, if spores were allowed to germinate this improved their retention. Therefore, further work was done to try and encourage spore germination.

Calcium stimulates metabolic and physiological processes in fungi (Pitt and Ugalde, 1984; Donaldson and Deacon, 1992), therefore, tests were done to examine the effect of adding calcium, as $CaCl_2$, at different concentrations to the overnight incubation solution. Calcium had different effects on retention, depending on the species of fungi. For *S. sclerotiorum*, the number of spores sticking to the plate was increased without improvement in the germination rate (Fig. 34.2). With *B. cinerea* and *A. brassicae*, calcium improved the germination rate, resulting in better retention of spores (Fig. 34.3, for *Botrytis cinerea*). Best results were obtained with calcium concentrations between 1 and 5 mM. The germination rate of all fungi tested was also enhanced by the addition of D-glucose to the incubation solution.

The background optical density readings were found to be increased by endogenous, substrate-relevant enzyme activity, which differed with fungal species, and substantially

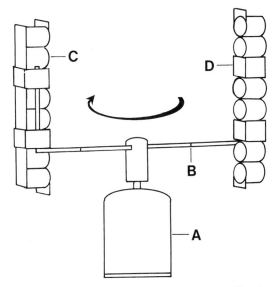

Fig 34.1. Schematic representation of the modified roto-rod spore trap. Direction of rotation is indicated by arrows. The arms rotate at about 2800 rpm and spores are impacted on the bottoms of the ELISA wells. A, 12 v electric motor; B, support arms; C, ELISA strips; D, rubber bands holding strips to support arms.

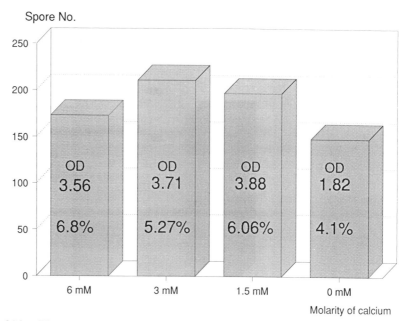

Fig 34.2. Effect of calcium on retention of *Sclerotinia sclerotiorum* ascospores. The number of ascospores retained in the wells after ELISA for different calcium concentrations. The ascospore germination rate and the relative optical density of the ELISA reading are also shown on the histogram bars. Results were obtained using monoclonal antibody 17/3/a.

decreased the sensitivity of the tests (Fig. 34.4). Various techniques have been suggested for reducing endogenous substrate-relevant enzyme activity (Cordell *et al.*, 1984; Ponder and Wilkinson, 1991), e.g. it has been reported to be inhibited by acetic acid (Ponder and Wilkinson, 1991), although the acid may destroy labile antigens. Experiments on the sensitivity of the epitopes recognized by the above antibodies showed acetic acid had a negligible effect on activity. However, when acetic acid was applied as an 15% aqueous solution for 3.5 min (after overnight incubation), endogenous substrate-relevant enzyme activity for spores or mycelium was reduced to a very low level for all fungi tested (Fig. 34.4).

PROTOCOL PRELIMINARY TEST FOR AIRBORNE FUNGAL SPORES

Materials

1. Nunc-Immuno-Modules, PolySorp F8 strips (Nunc, Multiwell polystyrene ELISA plates).
2. Modified roto-rod arms spore trap and power supply (Fig. 34.1).
3. Calcium chloride.
4. D-Glucose.
6. Distilled water.
7. Acetic acid.
8. Washing buffer – phosphate-buffered saline with 0.05% Tween-20.
9. Blocking solution – phosphate-buffered saline with 1% Nido milkpowder.

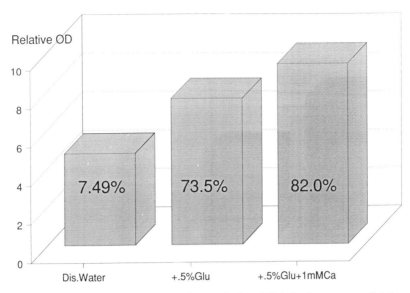

Fig 34.3. Effect of calcium and glucose on the germination of *Botrytis cinerea* spores. Relative optical density of the ELISA reading for different treatments. Spore germination rates are shown on the histogram bars. Results were obtained using monoclonal antibody BC-KH$_4$.

10. Rabbit antimouse immunoglobulins alkaline phosphatase conjugate (Sigma, Cat. No. D 314).

11. Substrate buffer – diethanolamine (97 ml) + MgCl$_2$ 6H$_2$O (100 mg) + 800 ml distilled water, adjusted to 9.8 pH with HCl and made up to 1 litre.

12. *p*-Nitrophenyl phosphate substrate tablets.

Method

The results of our experiments suggest that the following protocol could be used to detect airborne fungal spores of *A. brassicae*, *S. sclerotiorum* and *B. cinerea*. Further development and field tests will be needed to improve the technique.

1. Trap spores in ELISA strips using the modified roto-rod spore trap.
2. Remove strips from the spore trap and incubate overnight in distilled water containing D-glucose (0.5%) and calcium chloride (1 mM).
3. Wash the strips in washing buffer (phosphate-buffered saline with 0.05% Tween-20) three times, 3 min for each wash.
4. Apply acetic acid as a 15% aqueous solution for 3.5 min at room temperature.
5. Wash in washing buffer three times, 3 min for each wash.
6. Incubate in blocker solution (phosphate-buffered saline with 1% Nido milkpowder) for 30 min.
7. Wash in washing buffer three times, 3 min for each wash.
8. Apply specific monoclonal antibody in phosphate-buffered saline with 0.05% Tween-20 and 1% Nido milkpowder for 1 h at 37°C.
9. Wash in washing buffer three times, 3 min for each wash.
10. Apply rabbit antimouse antibody at 1/1000 dilution in phosphate-buffered saline with 0.05% Tween-20 and 1% Nido for 1 h at 37°C.
11. Wash in washing buffer three times, 3 min for each wash.

12. Apply substrate in substrate buffer at two tablets per 10 ml buffer.

13. Read optical density at selected time points.

LIMITATIONS

The response to the addition of calcium to improve spore retention was different for the fungal species tested. Therefore, for other systems the effectiveness of calcium in enhancing spore retention or germination needs to be investigated.

For the monoclonal antibodies tested here the use of acetic acid to reduce endogenous substrate-relevant enzyme activity had little effect on the antibody activity. However, as acetic acid may destroy labile antigens its use with other antibodies must be tested.

Future aims

The methods suggested above to detect airborne fungal pathogen inoculum are still at the experimental stage. Further development is needed before a practical system will be available for field use. Future work will include testing the above protocol in the field by comparing spore concentrations in oilseed rape plots measured conventionally and by this method. It is hoped that the technique will eventually be modified so that spore concentrations can be quantified using 'rot-rod dipstick assays'.

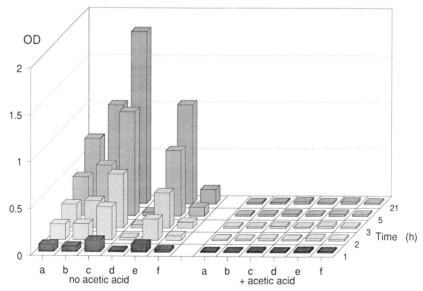

Fig 34.4. Endogenous, substrate-relevant enzyme activity for a range of fungal pathogens with and without application of 15% aqueous acetic acid for 3.5 min. Background optical density (405 nm) at different times. a, *Alternaria brassicae*; b, *Alternaria brassicicola*; c, *Alternaria alternata*; d, *Alternaria linicola*; e, *Botrytis cinerea*; f, *Alternaria infectoria*.

Availability of monoclonal antibodies

Workers interested in using monoclonal antibodies against *Alternaria brassicae* and/or *Sclerotinia sclerotiorum* should get in contact with one of the authors.

ACKNOWLEDGEMENTS

The authors are grateful to Dr Molly Dewey, Department of Plant Sciences, Oxford University for the supply of monoclonal antibodies to *B. cinerea*.

REFERENCES

Bartlett, J. T. and Bainbridge, A. (1978) Volumetric sampling of micro-organisms in the atmosphere. In: Scott, P. R. and Bainbridge, A. (eds) *Plant Disease Epidemiology*. Blackwell Scientific, Oxford.

Cordell, J. L., Falini, B., Erber, W. N., Ghosh, A. K., Abdulaziz, Z., MacDonald, S., Pulford, K. A. F., Stein, H. and Mason, D. Y. (1984) Immunoenzymatic labelling of monoclonal antibodies using immune complexes of alkaline phosphatase and monoclonal anti-alkaline phosphatase (APAAP complexes). *Journal of Histochemistry and Cytochemistry* 32(2), 219–229.

Donaldson, S. P. and Deacon, J. W. (1992) Role of calcium in adhesion and germination of zoospore cysts of *Pythium*: a model to explain infection of host plants. *Journal of General Microbiology* 138, 2051–2059.

Heusser, C. H., Stocker, J. W. and Gisler, R. H. (1981) Methods for binding cells to plastic: application to solid phase immunoassays for cell-surface antigens. *Methods in Enzymology* 73, 406–418.

Huang, W. M., Gibson, S. J., Facer, P., Gu, J. and Polak, J. M. (1983) Improved section adhesion for immunocytochemistry using high molecular weight polymers of L-lysine as a slide coating. *Histochemistry* 77, 275–279.

Mazia, D., Schatten, G. and Sale, W. (1975) Adhesion of cells to surfaces coated with polylysine. *Journal of Cell Biology*, 66, 198–200.

Pitt, D. and Ugalde, U. O. (1984) Calcium in fungi. *Plant, Cell and Environment* 7, 467–475.

Ponder, B. A. and Wilkinson, M. M. (1981) Inhibition of endogenous tissue alkaline phosphatase with the use of alkaline phosphatase conjugates in immunocytochemistry. *Journal of Histochemistry and Cytochemistry* 29(8), 981–984.

Index

Abbreviations:
    AFLP, Amplified fragment length
      polymorphism;
    ELISA, Enzyme-linked immunosorbent
      assay;
    ITS, Intragenic transcribed spacer;
      PCR, Polymerase chain reaction;
    RAPD, Random amplified polymorphic
      DNA;
    RFLP, Restriction fragment length
      polymorphism.

Absidia corymbifera 157
Acetic acid 250
African *Armillaria, see Armillaria*
Airborne spores 247–253
    analysis by indirect ELISA,
      calcium improving retention of 249,
        250, 252
      endogenous substrate-relevant enzyme
        reduction 249–250, 252
      limitations 252
      preliminary testing 250–252
      test development 248–250
    detection 247–253
    germination 249
    trapping methods 248
Alkaline phosphatase system 248–249
Allele-specific oligonucleotide (ASO)
      analysis 240–245
    limitations 243

probes 241–242
    see also under Venturia inaequalis
Alternaria, oilseed rape infection 247
Alternaria brassicae
    airborne spore detection 248–253
    monoclonal antibodies to spores 248,
      253
    spore germination and calcium
      effect 249
Alternaria dianthi 58, 61
Amplified fragment length polymorphisms
    (AFLPs) 76
    Fusarium oxysporum f.sp. *ciceris* 75–82
    procedure 76–78
Anthracnose (blackspot) 183
    see also individual Colletotrichum species
Antigens, fungal x
    sources and extraction x
    see also individual fungi
Apple scab 239
    see also Venturia inaequalis
Armillaria x, 141–156
Armillaria, African 141–147
    isolation in culture 142
    pairing test method 141
    RAPD analysis 141–142
      protocol 143–144
      results and limitations 145
    RFLP analysis 142
      protocol 144
      results and limitations 146

Armillaria, European 141, 149–156
  antigen extraction x, 155
  dipstick assay 153–154, 155
  ELISA using polyclonal antisera 150
  monoclonal antibodies 150, 154
  monoclonal antibody-based ELISA 150,
    151–153, 155
    protocol 151–153
    results and limitations 154–155
  pairing test methods 149, 154–155
Armillaria mellea 141, 149, 150
  monoclonal antibodies 150
Armillaria ostoyae 149, 150
  monoclonal antibodies 150
Ascochyta rabiei 78, 80
Aspergillus, monoclonal antibodies 158
Aspergillus niger 61, 166
  benomyl resistance 239

Bacteriophage M13 122, 124
Baiting and bait plants
  Rhizoctonia solani detection 29
  Spongospora subterranea contamination
    detection 23, 24
Bananas
  pathogenic fungal species 219
  Sigatoka leaf spots,
    DNA extraction from leaves 217–218,
    220
  see also Mycosphaerella species
Barley
  Gaeumannomyces–Phialophora complex
    infection 127
  Pyrenophora infections 191
Benomyl 239
  resistance,
    in *Aspergillus niger* and *Neurospora
    crassa* 239
  see also Venturia inequalis
Biological amplification 32
Black Sigatoka 215
Blackleg disease 1
Blackspot, *see* Strawberry blackspot disease;
    Anthracnose (blackspot)
Botryodiplodia theobromae 219
Botrytis, monoclonal antibodies 158
Botrytis aclada 174
Botrytis cinerea
  conidia 173
    collection 176

immunofluorescent detection using
    monoclonal antibodies 175–177
  detection in flowers 173–178
    conidia by monoclonal
      antibodies 173–178
    current plating methods 173
    polyclonal antibody-based
      methods 173
  detection in fruits/vegetables 165–172
    KH4 antigen location 167, 168
    monoclonal antibodies 166, 167, 248
      availability 170, 177
      to conidia 174, 177
      preparation 167, 174
      specificity 166
    monoclonal-antibody based ELISA,
      protocol 167–169
      results and limitations 169–170
    oilseed rape infection 247
    polyclonal antibody assays 165–166
    spore germination and calcium
      effect 249
Botrytis elliptica 174
Botrytis squamosa 174
Botrytis tullipae 174
Bread mould 157
Brown rot 165
Butt rot disease 111

Calcium, airborne spore retention in ELISA
    test 249, 250, 252
'Capture reagent' xii
Carbendazim 239
Carbohydrate extracts, high molecular
    weight x
  Mucor racemosus 158, 160
Carnations, *Fusarium oxysporum*
    detection 57–62
Cereals
  Fusarium culmorum and *F. graminearum*
    infections 37
  Pseudocercosporella herpotrichoides
    infection 9–15, 17–22
  take-all in 127
Chemiluminescence
  Fusarium culmorum detection 39–40
  Gaeumannomyces–Phialophora complex
    DNA detection 130–131
  Leptosphaeria maculans detection 5–6

Chickpea
  Fusarium wilt 75, 81
  yellowing 75, 81
Choanephora 157
Chrysanthemums, *Verticillium dahliae* DAS-
    ELISA test on 102
Cochliobulus sativum 196
Coffee berry disease 183
Colletotrichum 183
  culture and DNA extraction 184–185
  PCR amplification 185–186, 186–188
Colletotrichum acutatum 179–182
  C. gloeosporioides differentiation 179, 181,
      184
  current diagnostic tests 179
  DNA extraction 185
  immunogen preparation 180
  latent infection detection 179, 182
  monoclonal antibodies,
    preparation 180
    specificity 181
  monoclonal antibody-based
      ELISA 180–181, 182
  paraquat test 179, 182
  PCR amplification 187–188
    DNA purification step 188
  strawberry blackspot disease from 179,
      183
Colletotrichum coccodes 184
  PCR amplification 185–186, 188
Colletotrichum fragariae 179, 183, 188
Colletotrichum gloeosporioides 179, 181, 183
  Mycosphaerella PCR and 219
  PCR amplification 185, 186–187, 188
  sensitivity of PCR detection xi, 187, 188
Colletotrichum kahawe (*C. coffeanum*) 183
Colletotrichum lindemuthianum x, 235
Colletotrichum musae 219
Coniferous trees, *Heterobasidion annosum*
      infection 105, 111, 117
COST-88 programme ix
Cyclamen persicum, Fusarium oxysporum
      detection 57–62
Cyclophosphamide 109
Cylindrocarpon destructans 58, 61

Damping-off 29
DAS-ELISA (double antibody sandwich
      ELISA)
  Pseudocercosporella herpotrichoides 10, 13–15
  Pyrenophora graminea 201

Verticillium dahliae on
    chrysanthemums 102
Verticillium in roses 99–104
'Detection reagent' xii
Diaporthe phaseolorum f.sp. *meridionalis, see*
      Phomopsis phaseoli
Digoxigenin
  Fusarium culmorum DNA probe
      labelling 39
  Gaeumannomyces–Phialophora complex
      DNA detection 130–131
  Leptosphaeria maculans detection 5, 7
DNA
  amplification, *see* Polymerase chain
      reaction (PCR)
  dilution and RAPD-PCR limitation 114
  extraction,
    African *Armillaria* 142–143, 144
    Colletotrichum 184–185
    Fusarium culmorum and *F.
      graminearum* 39, 49
    Fusarium oxysporum f.sp. *gladioli* 64
    Gaeumannomyces–Phialophora
      complex 129
    Heterobasidion annosum 112, 118–119,
      123
    Leptosphaeria maculans 4
    Mycosphaerella species 217–218
    Phytophthora fragariae var. *rubi* 136–137
    Phytophthora infestans 226
    Pseudocercosporella herpotrichoides 19
    Pyrenophora graminea and *P. teres*
      192–193
    Sigatoka leaf spot (*Mycosphaerella*)
      infected plants 217–218, 220
    sodium dodecyl sulphate (SDS) 93
    Venturia inaequalis 241
    Verticillium albo-atrum and *V. dahliae* 93
    Verticillium from potatoes 85–87
  hybridization, *see* Hybridization
  mitochondrial, *see* Mitochondrial DNA
      (mtDNA)
  random amplified polymorphic, *see*
      Random amplified polymorphic
      DNA (RAPD)
  ribosomal, *see* Ribosomal DNA
  silver staining 121
DNA fingerprinting, *Phytophthora
      infestans* 225–229
  see also Phytophthora infestans

DNA probes
    Fusarium culmorum 38–46
        differentiation from *F.*
            graminearum 37–46
    Gaeumannomyces–Phialophora complex
        identification 127–134
        see also Gaeumannomyces–Phialophora
            complex
    Phytophthora fragariae var. *rubi* 135
    R-type *Pseudocercosporella*
        herpotrichoides 17–22
Dot-blot analysis
    Heterobasidion annosum detection 107, 108
    Pyrenophora graminea and *P. teres* 194–195
    see also Southern blotting
Dot-ELISA, *Pyrenophora graminea*
        detection 201–202, 203
Double antibody sandwich ELISA, *see* DAS-
        ELISA
Drechslera graminea, see Pyrenophora graminea

*Eco*RI 18, 19, 39
Electrophoresis
    Fusarium culmorum DNA 39–40
    Fusarium oxysporum detection in
        cyclamen/carnations 57–62
    Gaeumannomyces–Phialophora complex
        DNA 129–130
    Heterobasidion annosum DNA 123–124
        intragenic transcribed spacers
            (ITS) 121
    Leptosphaeria maculans DNA 4
    methods for fungi using isoenzymes 58
Electrorotation assay x–xi
ELISA
    Armillaria, European,
        monoclonal antibodies 150, 151–153,
            155
        polyclonal antisera 150
    Botrytis cinerea 165–172
    Erysiphe pisi 233–235
    Fusarium 48
    Heterobasidion annosum 106, 108
    indirect,
        airborne spore analysis, *see* Airborne
            spores
        Colletotrichum acutatum 180–181, 182
        Mucor racemosus 159
        spore retention improvement 249
    Mucor racemosus, see Mucor racemosus

Pseudocercosporella herpotrichoides 11
Pyrenophora graminea 201–202, 203
Pythium aphanidermatum 233–235
Rhizoctonia solani 30, 31–32
Spongospora subterranea 23–27
    method 24, 25
Verticillium 84
see also under individual fungi
Enzyme-linked immunosorbent assay, *see*
        ELISA
Enzymes, fungal x
    endogenous substrate-relevant in
        ELISA, reduction 249–250, 252
    Fusarium oxysporum species in
        cyclamen/carnation 57–62
    Phytophthora infestans markers 225
Epicoccum purpurascus 196
Erwinia carotovora 25, 58, 61
Erwinia chrysanthemi 58, 61
Erysiphe cichoracearum 235
Erysiphe graminis 235
Erysiphe pisi, monoclonal antibodies 231–238
    ELISA 233–235
    indirect immunofluorescence 232–233,
        237
    production and screening 232
    specificity 235–237
    UB7 236–237
Escherichia coli
    Pseudocercosporella herpotrichoides R-type
        DNA probe 18
    Pyrenophora cloning 193
α-Esterase, *Fusarium oxysporum* detection in
        cyclamen/carnation 57–62
European Community, COST-88 programme
        ix
Eyespot disease 9, 17
    see also Pseudocercosporella herpotrichoides

Flowers, *Botrytis cinerea* infection, *see Botrytis*
        cinerea
Fluorescence, *see* Immunofluorescence
Foliar diseases
    Erysiphe pisi 231–238
    Phytophthora infestans 223–230
    Pythium aphanidermatum 231–238
    Sigatoka leaf spots 215–221
    see also individual fungi
Fruit, *Botrytis cinerea* detection 165–172

Fungicide 9
    Pyrenophora graminea infections 199
    resistance,
        detection 239–245
        Phytophthora infestans 224
        see also Benomyl; *Venturia inaequalis*
Fusarium 9, 37, 47, 57
    ELISA-based assays 48
    genetic variation detection 48, 63
    taxonomy 37, 47
Fusarium avenaceum 9, 43, 44
Fusarium culmorum 9, 37, 47
    DNA isolation 39
    F. graminearum differentiation 37–46,
        47–55
    infections/infection route 37, 47
    RAPD-PCR 47–56
      assay technique 48–49
      assessment and primer
        screening 50–51
      limitations and perspectives 53–54
      primer UBC90 in 52
      profiles 51, 52
      species differentiation and
        genotypes 51–53
    RFLP probes 37, 38, 42
    SAGW29 strain 38, 39, 40, 42
    species-specific DNA probes 38–45
      isolation 38
      limitations 45
      shotgun cloning of Sau3A
        fragments 38
Fusarium eumartii 78, 80
Fusarium graminearum 37, 47
    F. culmorum differentiation 37–46, 47–55
    infections/infection route 37, 47
    RAPD-PCR 47–56
      profiles 51, 52
    strain DSM62223 reclassified 42
    see also Fusarium culmorum
Fusarium moniliforme 205
    Mycosphaerella differentiation 219
    PCR amplification of genomic
      DNA 207
    RAPD 207, 210, 212
Fusarium oxysporum 44, 57, 63
    electrophoretic detection in cyclamen/
      carnations 57–62
    internal control templates for PCR of
      Verticillium 87

Fusarium oxysporum f.sp. *ciceris* 75–82
    amplified fragment length
      polymorphisms (AFLPs) 75–82
    protocol 75–78
    results 78–80
    differentiation from other chickpea
      pathogens 78, 81
    infection route and symptoms 75
    races 75–76
Fusarium oxysporum f.sp. *conglutinans* 76
Fusarium oxysporum f.sp. *cyclaminis* 57–62
Fusarium oxysporum f.sp. *dianthi* 57–62
Fusarium oxysporum f.sp. *gladioli* 63–68
    laboratory test for latent infection
      detection 63
    origin 65
    races 63–64
    RFLP and RAPD analysis 63–68
      protocol 64–65
      results and limitations 65–67
    vegetative compatibility groups
      (VCG) 64, 65
Fusarium oxysporum f.sp. *mathioli* 76
Fusarium oxysporum f.sp. *melonis* 76
Fusarium oxysporum f.sp. *niveum* 76
Fusarium oxysporum f.sp. *pisi* 69–74
    cultivars resistant to 70
    genetic variants (races) 69–70
    infection route and symptoms 69
    race 6 strains from Denmark 72, 73
    RFLPs 69–74
      procedures 70–71
      results and limitations 71–73
Fusarium oxysporum f.sp. *raphani* 76
Fusarium redolens 57
Fusarium solani 78, 80
Fusarium sulphureum 45
Fusarium wilt 69, 75
Fusarium yellows 63

Gaeumannomyces graminis, PCR assay 133
Gaeumannomyces graminis var. *avenae* 127
Gaeumannomyces graminis var. *tritici* 127
Gaeumannomyces–Phialophora complex 127
    DNA extraction 129
    identification by DNA probes 127–134
      GggMR1/pEG34 128, 131–132,
      132–133
      pMSU315 128, 131, 132
      protocol 128–131

Gaeumannomyces–Phialophora complex *contd*
  results and limitations 131–133
  identification by PCR procedures 130,
    133
α-D-Galactosidase 59, 60
β-D-Galactosidase 59
Genotyping, *Fusarium culmorum* 51–53
Gibberella fujikuroi, see Fusarium moniliforme
Gladioli
  corm rot 63
  Fusarium oxysporum f.sp. *gladioli*
    infection 63
  'large-flowered' 63
Gliocladium roseum 181
Glomerella cingulata (*Colletotrichum*
    gloeosporioides) 179, 181
Glomus caledonium 58
Glomus mosseae 58
β-D-Glucosidase
  electrophoretic methods 58
  Fusarium oxysporum detection in
    cyclamen/carnation 57, 59, 60
Glutamate oxaloacetate transaminase 59, 60
Grapes, *Botrytis cinerea* detection 167–172
Grapevine, grey mould 165

Heartrot, *Heterobasidion annosum* causing 105
Heterobasidion annosum x, 105–110, 111–116,
    117–126
  classification basis 117
  current detection methods 105
  DNA extraction 112, 118–119, 123
  intersterility groups (strains) 105, 111,
    117
  host range 115, 117
  identification by intragenic transcribed
    spacers 118, 120
  identification by RAPD-PCR
    111–116, 118
  identification by rDNA gene
    polymorphism 117–126
  RFLP analysis 118
  see also techniques below
  minisatellite allele polymorphisms 118,
    122–124
  monoclonal-based assays,
    antibody specificity 107
  limitations 108–109
  procedures 106–107
  reasons for need 105–106

polyclonal antisera assays 106
RAPD-PCR 111–116, 118
  primers 112
  procedure 111–114
  results and limitations 114–115
  ribosomal DNA gene
    polymorphism 117–126
  differential PCR amplification 122
  intragenic transcribed spacers
    (ITS) 118, 120
  ITS amplification and analysis
    protocol 118–122
  limitations 124–125
  single DNA strand conformational
    polymorphism 120–121, 122
Hexadecyl trimethylammonium bromide
    (CTAB) 86, 87
Hops, *Verticillium albo-atrum* infection 91
Howard Mould Count (HMC) 158
Hybridization
  Fusarium culmorum 38–45
  Fusarium oxysporum f.sp. *pisi* 71
  Heterobasidion annosum 123–124
  Leptosphaeria maculans DNA 2, 4–5
  Pseudocercosporella herpotrichoides
    R-type 18–19
  see also Southern blotting; *other specific*
    techniques
Hybridoma technology ix–x, 167
  see also Monoclonal antibodies

Immunoassays
  double antibody sandwich,
    Pseudocercosporella herpotrichoides 10,
    13–15
  for field-use xi
  see also specific techniques
Immunofluorescence
  Botrytis cinerea conidia detection 175–177
  colony staining assay for *Rhizoctonia*
    solani 30, 32–34
  indirect, *Pythium aphanidermatum* and
    Erysiphe pisi 232–233, 237
  Pyrenophora graminea detection 200–201,
    202
Immunofluorescent colony staining assay
    (IFC), *Rhizoctonia solani* 30, 32–34
In vitro technology xi

Indirect immunofluorescence, *Pythium aphanidermatum* and *Erysiphe pisi* 232–233, 237
Internal (intragenic) transcribed spacers (ITS) xi
  Colletotrichum species xi, 185–186, 188
  Heterobasidion annosum 118
    amplification 120–121
    analysis 122
    limitations 124–125
  Verticillium albo-atrum and *V. dahliae* subgroup detection 92
  Verticillium analysis xi, 84, 92
Isoenzymes, *Fusarium oxysporum* detection in cyclamen/carnation 57–62

Leaf spot disease, Sigatoka 215
Leaf stripe disease 191, 199, 205
Leptosphaeria maculans,
  aggressive/non-aggressive strain differentiation 1
  classification based on RAPD-PCR 2
  DNA preparation 4
  DNA probes (A- and NA-) 7
  RAPD-PCR diagnosis 1–8
    protocol 2–6
    results 6–7
Lucerne, *Verticillium albo-atrum* infection 91

Magnesium chloride 114
Maize, stalk and cob rot 205
Microdochium nivale 9, 12
Micronectriella nivalis 45
  Fusarium RAPD-PCR screening 51
Minisatellite allele polymorphisms, *Heterobasidion annosum* 118, 122–124
Minisatellite elements 122
  probes 122, 124
Mitochondrial DNA (mtDNA) 76
  Phytophthora infestans 224
  RFLPs 76
Monoclonal antibodies ix–x
  airborne spores 248
  Alternaria brassicae 248, 253
  Armillaria 150–151, 154
  Aspergillus 158
  Botrytis 158
  Botrytis cinerea 166, 167, 248
    conidia 174

  Colletotrichum acutatum 180, 181–182
  Erysiphe pisi 231–238
  Heterobasidion annosum 106, 107
  Penicillium 158
  Pseudocercosporella herpotrichoides 9–15, 10–12
  Pyrenophora graminea 199–200
  Pythium aphanidermatum 231–238
  Rhizoctonia solani 29, 30–31
  Sclerotinia sclerotiorum 248, 253
  Spongospora subterranea 23–27
  Verticillium dahliae 100
  see also individual fungi
Monoclonal antibody-based ELISA, *see* ELISA
Mortierella 161, 162
Moulds, in food 157–158
Mucor circinelloides 161
Mucor racemosus 157–164
  dot-blot assays 159–160
  ELISA using polyclonal antisera 158
  extracellular polysaccharides (EPS) 158
    immunoreactivity of epitopes 162
    preparation 158–159
  indirect ELISA using monoclonal antibodies 158–64, 159
    limitations 161–163
    protocol 159
    results 161–162
  monoclonal antibodies 160, 161, 162–163
    preparation 158
Mucorales 157
  ELISA and dot-blot assay 157–164
  see also Mucor racemosus
Mucormycosis, detection 163
Mutations, fungicide resistance 239–240
  detection by allele-specific oligonucleotide (ASO) analysis 240–245
Mycosphaerella fijiensis 215
  black Sigatoka 215
  DNA extraction 217–218
  PCR 215–221
    protocol 218
    results and limitations 219–220
  ribosomal DNA internal transcribed spacers (ITS) 216
  spore isolation 218
Mycosphaerella minima 219

Mycosphaerella musicola 215
    DNA extraction 217–218
    PCR 215–221
      protocol 218
      results and limitations 219–220
      rDNA internal transcribed spacers
        (ITS) 216
      spore isolation 218
      yellow Sigatoka from 215
Mycotoxins 37, 47

Net blotch and leaf spot disease 191, 199,
    205
Neurospora crassa, benomyl-resistance 239
    Codon 198 mutation 240

Oats, *Gaeumannomyces–Phialophora* complex
    infection 127
Oilseed rape
    airborne spore detection 247–253
    fungal pathogens 247
    Leptosphaeria maculans diagnosis 1–8
Oligonucleotide fingerprinting, *see* DNA
    fingerprinting
Oospores 223, 229

Paracercospora fijiensis, see Mycosphaerella fijiensis
Paraquat test, *Colletotrichum acutatum* 179, 182
Partially purified proteins x, 174
Pea
    cultivars resistant to *Fusarium oxysporum*
      f.sp. *pisi* 70
    Fusarium wilt 69
    powdery mildew 235
Pectin lyase 59, 60
Penicillium 34
    monoclonal antibodies 158
Peronospora pisi 235
Peroxidase 59, 60
Phialophora cinerescens 58, 61
Phialophora species 127
Phoma lingam 247
Phomopsis phaseoli f.sp. *meridionalis* 206
    PCR amplification of genomic
      DNA 208–209
    RAPD 208–209, 211, 212–213
Phytophthora 57, 135
Phytophthora fragariae var. *fragariae* 135

Phytophthora fragariae var. *rubi* 135
    detection by PCR 135–139
    DNA isolation 136–137
    DNA probes 135
    Southern blot hybridization 135, 138
Phytophthora infestans 223–230
    DNA fingerprinting 225–229
      protocol 226–227
      results 227–229
    DNA isolation 226
    double-stranded RNA 224
    fungicide resistance 224
    infection caused by 223
    markers 224–225
      allozyme 225
      biologically significant 224
      cytoplasmic 224–225
    mating types 223, 224
    mitochondrial DNA 224
    polymorphic DNA markers 225
    sexual reproduction 223, 229
    virulence 224
Pichia membranaefaciens 158, 160
Plantain, Sigatoka leaf spots 215–221
    see also Mycosphaerella species
Polygalacturonase 59, 60
Polymerase chain reaction (PCR) xi
    advantages 84
    arbitrarily primed (AP-PCR) 76
      see also RAPD-PCR
    Colletotrichum species 185–186, 186–188
    Fusarium culmorum detection 39
    Fusarium moniliforme DNA 207
    Gaeumannomyces graminis
      identification 133
    Gaeumannomyces–Phialophora complex
      DNA 130, 133
    inhibitors 88–89
      Colletotrichum acutatum detection 188
    Leptosphaeria maculans 2, 4, 6–7
    Mycosphaerella species 216–221
    Phomopsis phaseoli f.sp. *meridionalis*
      DNA 208–209
    Phytophthora fragariae var. *rubi* 135–139
    practical assay problems xi–xii
    practical guidelines 53
    primer development,
      Colletotrichum 185–186, 187
      Fusarium culmorum 50–51, 52
      Heterobasidion annosum 112
      Mycosphaerella 216, 217

Polymerase chain reaction (PCR) *contd*
    Verticillium 92–93, 94
    Pyrenophora DNA 206–207
    quantification of *Verticillium* xi, 88
    RAPD-PCR, *see* Random amplified
        polymorphic DNA (RAPD)
    ribosomal DNA,
        Gaeumannomyces–Phialophora complex
          DNA 133
        intragenic transcribed spacers (ITS) of
          Heterobasidion annosum 118, 122
        Mycosphaerella 216
    sensitivity xi, 138, 187, 188
    Stenocarpella maydis DNA 207
    technical problems xi
    technique, in *Verticillium* 85–88, 95–96
    Venturia inaequalis 241
    Verticillium albo-atrum 91–97
    Verticillium dahliae 91–97
    Verticillium species detection in
        potato 83–89
Postharvest diseases
    Botrytis cinerea 165–172, 173–178
    Colletotrichum 183–189
    Colletotrichum acutatum 179–182
    Mucor racemosus 157–164
    see also individual fungi
Potato early dying complex 83
Potato mop-top virus 23
Potatoes
    late blight 223
        see also Phytophthora infestans
    stem canker 29
    vascular wilt disease 83
    Verticillium detection by PCR 83–89
Powdery scab 23
Proteins, partially purified x, 174
Pseudocercospora musae, *see Mycosphaerella musicola*
Pseudocercosporella anguioides 17
Pseudocercosporella herpotrichoides x, 45
    antigen extraction x, 10
    antisera 10
        cross-reactions 9–10
        production 10–12
    detection in plant tissue 19
    monoclonal antibody-based
        immunoassay 9–15
        double antibody sandwich method 10,
          13–15

    limitations 14
    production 10–12
    R-type 17
    R-type DNA probe 17–22
        development 18
        results and limitations 19–21
    W-type 17
Pseudocercosporella herpotrichoides var. *acuformis* 51
Pyrenopeziza brassicae 247
Pyrenophora 205
    RAPD 206–207, 209–210
Pyrenophora bromi 196
Pyrenophora graminea 191, 199, 205
    culture 192
    DAS-ELISA (double antibody sandwich
        ELISA) 201
    DNA extraction and cloning 192–193
    dot-blot identification 191–197
        protocol 194–195
        results and limitations 195, 196
    dot-ELISA 201–202, 203
    immunofluorescence microscopy
        200–201, 202
    infection from 191, 199, 205
    monoclonal antibodies 199–200
        production 200
        specificity 200, 203
    P. teres differentiation,
        acid phosphatase isoenzyme
          analysis 191
        importance of 199
        monoclonal antibody-based
          methods 200–203
        pathogenicity test on barley 191
        RAPD 192, 206–207, 208–209
    PCR amplification of genomic
        DNA 206–207
    pigments from 191
    plasmid DNA 193, 195
    RAPD 192, 206–207, 208–209
    RFLP identification 191–197
        protocol 193–194
        results 195–196
Pyrenophora teres 191, 199, 205
    DNA extraction and cloning 192–193
    infection from 191, 199, 205
    P. graminea differentiation, *see Pyrenophora
        graminea*
    PCR amplification of genomic
        DNA 206–207

Pyrenophora teres contd
    pigments from 191
    plasmid DNA 193, 195
    RAPD 206–207, 208–209
Pythium aphanidermatum, monoclonal
        antibodies 231–238
    characteristics 234
    ELISA 233–235
    indirect immunofluorescence 232–233,
        237
    production and screening 232
    specificity 234, 235–237

Quantitation of fungal infection,
        Verticillium xi, 88

Random amplified polymorphic DNA
        (RAPD), *see* RAPD-PCR
RAPD 2
RAPD-PCR 2, 48, 76
    African *Armillaria* 141–142, 143–144,
        145
    assay technique 2, 4–5, 48–49, 50
    Fusarium colmorum and *F.
        graminearum* 47–56
    Fusarium moniliforme 207, 210, 212
    Fusarium oxysporum f.sp. *gladioli*
        analysis 64, 66
    guidelines and protocol requirements 53
    Heterobasidion annosum 111–116, 118
    Leptosphaeria maculans 1–8
    limitations 114
    Phomopsis phaseoli f.sp. *meridionalis*
        208–209, 211, 212–213
    Phytophthora infestans DNA 225
    Pyrenophora 206–207, 209–210
    Pyrenophora graminea 192, 206–207,
        208–209
    Stenocarpella maydis 207, 210, 212
    see also Polymerase chain reaction (PCR);
        under individual fungi
Rapeseed, *see* Oilseed rape
Raspberry, *Phytophthora fragariae* var. *rubi*
    detection 135–139
rDNA sequences, *see* Ribosomal DNA
Restriction endonuclease digestion
    Gaeumannomyces–Phialophora complex
        DNA 129–130
    Heterobasidion annosum DNA 123
    Phytophthora fragariae var. *rubi* 137

    *see also Eco*RI
Restriction fragment length polymorphism
        (RFLP) xi
    African *Armillaria* 142, 144, 146
    Fusarium culmorum differentiation from *F.
        graminearum* 37–46
    Fusarium oxysporum f.sp. *gladioli*
        analysis 64, 65
    Fusarium oxysporum f.sp. *pisi* 69–74
    Heterobasidion annosum intersterility
        groups 118
    mitochondrial DNA, *Fusarium
        oxysporum* 76
    probes,
        Fusarium culmorum 37, 42
        Fusarium oxysporum f.sp. *pisi* 71
    Pyrenophora graminea and *P. teres* 193–194,
        195
    Verticillium albo-atrum and *V.
        dahliae* 91–92
RFLP, *see* Restriction fragment length
        polymorphism (RFLP)
Rhizoctonia brown patch, detection 30
Rhizoctonia cerealis 9, 10, 12, 34, 45
Rhizoctonia solani 12, 29
    antigen extraction 31
    antisera and cross-reactions of 29–30
    monoclonal antibodies production 29,
        30–31
    monoclonal antibody-based
        immunoassays x, 29–35
    diagnostic-ELISA test 30, 31–32
    immunofluorescent colony staining
        assay (IFC) 30, 32–34
    protocols 31–32
    results 32
    semi-selective medium for 32, 34
Rhizomucor pusillus 161
Rhizopus oligosporus 157
Rhizopus stolonifer 157
Ribosomal DNA (rDNA) xi, 135
    Colletotrichum species xi, 185–186, 188
    Heterobasidion annosum intersterility
        groups, *see Heterobasidion annosum*
    ITS, *see* Internal (intragenic) transcribed
        spacers (ITS)
    Mycosphaerella species 216
    PCR-amplified, *Gaeumannomyces–
        Phialophora* complex identification 133

Ribosomal genes
    PCR for *Verticillium* 84
    PCR for *Verticillium albo-atrum* and
        V. dahliae 92–96
    see also rDNA; rRNA
Ribosomal RNA (rRNA) 118
    PCR for *Verticillium* xi, 84, 92
    Verticillium albo-atrum and *V. dahliae*
        subgroup detection 92
RNA, double-stranded (dsRNA), *Phytophthora*
    infestans 224
Root discoloration 127
Root rot
    Armillaria causing 141, 149
    Heterobasidion annosum causing 105, 111
Root-infecting fungi
    Armillaria 141–147, 149–156
    Gaeumannomyces–Phialophora
        complex 127–134
    Heterobasidion annosum 105–110,
        111–116, 117–126
    Phytophthora fragariae var. *rubi* 135–139
    see also individual fungi
Roses, *Verticillium* detection by DAS-
    ELISA 99–104
rRNA, *see* Ribosomal RNA
Rye, *Pseudocercosporella herpotrichoides* R-type 17

Sclerotinia sclerotiorum 247
    monoclonal antibodies 248, 253
    spore retention for ELISA and calcium
        effect 249, 250
SDS extraction, of DNA 93
SDS polyacrylamide gel electrophoresis,
    Armillaria monoclonal antibody
    development 150
Seed-borne diseases
    Fusarium moniliforme 205–213
    Phomopsis/Diaporthe complex 205–213
    Pyrenophora 205–213
        Pyrenophora graminea 191–197, 199–203
        Pyrenophora teres 191–197
    Stenocarpella maydis 205–213
    see also individual fungi
Seedling blight 205
Septoria nodorum 45
'Sequence characterized amplified regions'
    (SCARs) 54
Shotgun cloning, *Fusarium culmorum* Sau3A
    fragments 38

Sigatoka leaf spots
    features of infections 215
    forms (black and yellow) 215
    see also Mycosphaerella species
Silicon carbide paper 13, 15
Single DNA strand conformational
    polymorphism (SSCP) 118, 122
    Heterobasidion annosum intersterility
        groups 120–121
    protocol 122
Soil, *Spongospora subterranea* contamination
    detection 23
Soil-borne diseases
    Fusarium culmorum 37–46, 47–56
    Fusarium graminearum 37–46, 47–56
    Rhizoctonia solani 29–35
    Spongospora subterranea (powdery
        scab) 23–27
    see also individual fungi
Southern blotting
    Fusarium culmorum DNA 39–40, 41, 45
    Fusarium oxysporum f.sp. *pisi* 70–71
    Gaeumannomyces–Phialophora complex
        DNA 129–130
    limitations 133
    Phytophthora fragariae var. *rubi* 135, 138
    Phytophthora infestans 227
    Pseudocercosporella herpotrichoides
        detection 20
    Pyrenophora graminea and *P. teres* 194–195
    see also Dot-blot analysis
Soybean stem canker, *see also Diaporthe*
    phaseolorum, 206
Spongospora subterranea 23–27
    monoclonal antibody production 24–25
    plasmodia 23, 24
    zoospores 23, 24
        ELISA detection 23–27
        limit of detection 25, 27
        preparation 24
        quantification 24, 27
Spores, *see* Airborne spores
Squash blot, procedure 93
Stalk and cob rot, of maize 205
Stem canker 29
Stem-based diseases
    Leptosphaeria maculans infections (blackleg
        disease) 1–8
    Pseudocercosporella herpotrichoides
        (eyespot) 9–15, 17–22
    see also individual fungi

Stenocarpella maydis 205
    PCR amplification of genomic
       DNA 207
    RAPD 207, 210, 212
Strawberry
    anthracnose, *see* Strawberry blackspot
      disease
    Botrytis cinerea detection 167–172
    Colletotrichum DNA extraction 185
Strawberry blackspot disease 179, 183
    see also Colletotrichum acutatum
Syncephalastrum racemosum 157

Take-all in cereals, *see also Gaeumannomyces–*
    Phialophora complex, 127
Taq polymerase 50, 51
Thamnidium elegans 157
Thielaviopsis basicola 58, 61
Tomato
    Colletotrichum DNA extraction 185
    late blight 223
Tomato plants, bait plants for *Spongospora*
    subterranea detection 23
Trichoderma viride 34
Tubers
    Spongospora subterranea infection 23
    Verticillium infection 83
β-Tubulin, in benomyl-resistant strains 239
    DNA amplification 241, 243
    mutations 239–240, 242–243

UPGMA cluster analysis 49, 52

Vascular pathogens
    Fusarium oxysporum 57–62
    Fusarium oxysporum f.sp. *ciceris* 75–82
    Fusarium oxysporum f.sp. *gladioli* 63–68
    Fusarium oxysporum f.sp. *pisi* 69–74
    Verticillium 83–89, 91–97, 99–104
    see also individual pathogens
Vascular wilt disease
    potatoes 83
    Verticillium 99
    Verticillium albo-atrum and *V. dahliae* 91
Vegetables
    anthracnose due to *Colletotrichum*
      coccodes 184
    Botrytis cinerea in 165

Venturia inaequalis 239
    allele-specific oligonucleotide (ASO)
      analysis 240–245
    limitations 243–245
    probes 241–242
    protocol 240–242
    results 242–243
    β-tubulin gene mutations 239–240,
      242–243
    benomyl-resistance 239
      Codon 198 mutation 239, 240
    DNA extraction and PCR 241
Venturia pirina 243
Verticillium
    DNA preparation 85–87
    ELISA test 84
    infection route and disease 83–84
    PCR-based assays 83–89
      internal control templates 87
      limitations and inhibitors 88–89
      protocol 85–88
      rationale/background to 84–85
    in roses, double-antibody sandwich
      ELISA 99–104
    standard methods of detection 83, 84
Verticillium albo-atrum 83, 99
    DNA extraction 93
    host-adapted haploids 91
    PCR-based assays 91–97
      limitations 96
      primer development 92–93, 94
    RFLP studies 91–92
    subgroups 84–85, 92
Verticillium dahliae 83, 99
    antigen preparation 100–101
    DNA extraction 93
    host-adapted haploids and diploids 91
    immunoassays,
      monoclonal antibody-based 100
      polyclonal antibody-based 100
    infection route and symptoms 99
    monoclonal antibody preparation 100
    PCR-based assays 91–97
      limitations 96
      primer development 92–93, 94
    RFLP studies 91–92
    in roses, detection by DAS-ELISA
      99–104
      limitations and results 101–103
      procedure 100–101
    subgroups 92

Verticillium nigrescens 100
Verticillium tricorpus 83
Verticillium wilt 83, 99–104

Western blotting
    Armillaria 150, 152
    Pseudocercosporella herpotrichoides 11

Wheat
    Gaeumannomyces–Phialophora complex
        infection 127
    Pseudocercosporella herpotrichoides W- and
        R-types 17
Wheat germ agglutinin 15

'Yellow pulp' 215
Yellow Sigatoka 215